SYMBOLIC COMPUTATION

Artificial Intelligence

Springer Series
SYMBOLIC COMPUTATION – *Artificial Intelligence*

N.J. Nilsson: Principles of Artificial Intelligence. XV, 476 pages, 139 figs., 1982

J.H. Siekmann, G. Wrightson (Eds.): Automation of Reasoning 1. Classical Papers on Computational Logic 1957–1966. XXII, 525 pages, 1983

J.H. Siekmann, G. Wrightson (Eds.): Automation of Reasoning 2. Classical Papers on Computational Logic 1967–1970. XXII, 638 pages, 1983

L. Bolc (Ed.): The Design of Interpreters, Compilers, and Editors for Augmented Transition Networks. XI, 214 pages, 72 figs., 1983

M.M. Botvinnik: Computers in Chess. Solving Inexact Search Problems. With contributions by A.I. Reznitsky, B.M. Stilman, M.A. Tsfasman, A.D. Yudin. Translated from the Russian by A.A. Brown. XIV, 158 pages, 48 figs., 1984

L. Bolc (Ed.): Natural Language Communication with Pictorial Information Systems. VII, 327 pages, 67 figs., 1984

R.S. Michalski, J.G. Carbonell, T.M. Mitchell (Eds.): Machine Learning. An Artificial Intelligence Approach. XI, 572 pages, 1984

A. Bundy (Ed.): Catalogue of Artificial Intelligence Tools. Second, revised edition. IV, 168 pages, 1986

C. Blume, W. Jakob: Programming Languages for Industrial Robots. XIII, 376 pages, 145 figs., 1986

J.W. Lloyd: Foundations of Logic Programming. Second, extended edition. XII, 212 pages, 1987

L. Bolc (Ed.): Computational Models of Learning. IX, 208 pages, 34 figs., 1987

L. Bolc (Ed.): Natural Language Parsing Systems. XVIII, 367 pages, 151 figs., 1987

N. Cercone, G. McCalla (Eds.): The Knowledge Frontier. Essays in the Representation of Knowledge. XXXV, 512 pages, 93 figs., 1987

G. Rayna: REDUCE. Software for Algebraic Computation. IX, 329 pages, 1987

D.D. McDonald, L. Bolc (Eds.): Natural Language Generation Systems. XI, 389 pages, 84 figs., 1988

L. Bolc, M.J. Coombs (Eds.): Expert System Applications. IX, 471 pages, 84 figs., 1988

C.-H. Tzeng: A Theory of Heuristic Information in Game-Tree Search. X, 107 pages, 22 figs., 1988.

Chun-Hung Tzeng

A Theory of
Heuristic Information
in Game-Tree Search

With 22 Figures

Springer-Verlag
Berlin Heidelberg New York
London Paris Tokyo

Chun-Hung Tzeng
Department of Computer Science
College of Sciences and Humanities
Ball State University
Muncie, IN 47306, USA

ISBN-13:978-3-642-64812-0   e-ISBN-13:978-3-642-61368-5
DOI: 10.1007/978-3-642-61368-5

Library of Congress Cataloging-in-Publication Data.
Tzeng, Chun-Hung.
A theory of heuristic information in game-tree search / by Chun-Hung Tzeng. p. cm. –
(Symbolic computation. Artificial intelligence)
Bibliography: p. Includes index.
ISBN-13:978-3-642-64812-0
1. Artificial intelligence. 2. Game theory. 3. Logic, Symbolic and mathematical. I. Title.
II. Series. Q335.T99 1988 519.3–dc 19 88-3105 CIP

© Springer-Verlag Berlin Heidelberg 1988
Softcover reprint of the hardcover 1st edition 1988
Typesetting: Macmillan India Ltd., Bangalore, India
2145/3140-543210

In Memory of My Parents

Mr. Peng-Yuo Tzeng
Mrs. Fu Cheng-Mei Tzeng

Preface

This book introduces a theoretical study, based on probabilistic game models, of heuristic information in game-tree searches. A precise formulation of heuristic information about a given game is introduced for a probabilistic game model, and heuristic search is formalized as a procedure for collecting heuristic information at a node in a game search tree. In the search tree (which is only a portion of the corresponding complete game tree), a heuristic search accumulates heuristic information at all of the nodes in the search tree. The more nodes the search tree has, the more precise is the heuristic information which the heuristic search collects.

In this book, a notion of node strength is also formulated to allow for the development of a mathematical theory for estimating node strength based on heuristic information. In this theory, the quality of decision making based on heuristic information is always improved if the corresponding heuristic information becomes more precise. Two specific game models are used for deriving actual procedures for estimating special node strength (i.e., minimax value). One method uses the product-propagation rules as the back-up procedure and the other uses no back-up procedure at all.

The use of heuristic information is crucial in many artificial intelligence (AI) systems and is usually based on intuition and experience. Since there are still many controversial issues about the usage of heuristic information, its study is important in the field of AI. No actual procedure for game-tree searches is proposed here. Rather, starting with probabilistic game models, this book is an attempt to build a theoretical foundation of heuristic information.

The author wishes to thank Dr Paul Purdom and Dr Cynthia Brown for their guidance and encouragement throughout his study at Indiana University. This book is based upon work supported by Ball State University and the National Science Foundation. The author wishes to thank both organizations for their support.

The author is grateful to Springer-Verlag for soliciting and publishing this book. The author also thanks the reviewer and Dr Loveland, Managing Editor, for their valuable suggestions. Special appreciation is extended to Ms Lucille Bailey for editing and Mrs Judith Bonneau for typing. Finally, the author wishes to thank his family, especially his brother Mr Chun-Ting Tzeng, for encouragement and moral support.

Muncie, Indiana, February 1988   *Chun-Hung Tzeng*

Acknowledgement

This material is based upon work supported by the National Science Foundation under Grant No. IST-8505735. Any opinions, findings, and conclusions or recommendations expressed in this publication are those of the author and do not necessarily reflect the views of the National Science Foundation.

Contents

1 **Introduction** 1

2 **Games and Minimax Values** 7

2.1 Finite Perfect Information Games and Game Trees 7
2.2 Zero-Sum Two-Person Perfect Information Games 8
2.3 Subgames and Game Graphs 9
2.4 Example 1: G_1-Games 10
2.5 Example 2: P_2-Games 11
2.6 Minimax Values, the Minimax Procedure and the Alpha-Beta
     Procedure 13

3 **Heuristic Game-Tree Searches** 18

3.1 The Conventional Heuristic Game-Tree Search 19
3.1.1 Static Evaluation Functions 19
3.1.2 The Back-Up Process 20
3.2 Heuristic Arguments and the Pathological Phenomenon 21
3.3 New Back-Up Processes 22
3.3.1 The Product-Propagation Procedure 23
3.3.2 The $M \& N$ Procedure 24
3.3.3 The *-MIN Procedure 25
3.3.4 Average Propagation 26

4 **Probability Spaces and Martingales** 28

4.1 Borel Fields and Partitions 28
4.2 Probability Spaces 31
4.3 Random Variables 33
4.4 Product Spaces 35
4.5 Conditional Probabilities and Martingales 36

5 **Probabilistic Game Models and Games Values** 40

5.1 Probabilistic Game Models 40
5.2 Strategies and Game Values 41
5.2.1 Non-randomized Strategies 42

5.2.2   Randomized Strategies   43
5.2.3   Minimax Values   44
5.3 P_b-Game Models   45
5.4 G_d-Game Models   46

6   **Heuristic Information**   51

6.1   Examples: P_2- and G_1-Game Models   52
6.1.1 P_2-Game Models   52
6.1.2 G_1-Game Models   56
6.2   Formulation of Heuristic Information   57
6.3   Heuristic Search   59
6.4   Improved Visibility of a Heuristic Search   61

7   **Estimation and Decision Making**   64

7.1   Random Variable Estimators   65
7.2   Comparison of Estimators   66
7.3   Decision Making   68
7.3.1   Decision Models   68
7.3.2   Decision Qualities   70

8   **Independence and Product-Propagation Rules**   73

8.1   Product Models   73
8.2   Product Heuristic Information and Product Heuristic Searches . .   75
8.3   Product-Propagation Rules   77

9   **Estimation of Minimax Values in P_b-Game Models**   81

9.1   More About Probabilities on P_b-Game Trees   81
9.2   The Conditional Probability of a Forced Win, $p(h,l)$ 82
9.3   An Approximation of $p(h,l)$ 86

10   **Estimation of Minimax Values in G_d-Game Models**   87

10.1   Estimation in G_1-Game Models   88
10.2   Estimation in G_d-Game Models   91

11   **Conclusions**   98

Appendix 100

References 102

Subject Index 104

1 Introduction

Searching is an important process in most AI systems, especially in those AI production systems consisting of a global database, a set of production rules, and a control system. Because of the intractability of uninformed search procedures, the use of heuristic information is necessary in most searching processes of AI systems. This important concept of heuristic information is the central topic of this book. We first use the 8-puzzle and the game tic-tac-toe (noughts and crosses) as examples to help our discussion.

The 8-puzzle consists of eight numbered movable tiles set in a 3×3 frame. One cell of the frame is empty so that it is possible to move an adjacent numbered tile into the empty cell. Given two tile configurations, initial and goal, an 8-puzzle problem consists of changing the initial configuration into the goal configuration, as illustrated in Fig. 1.1. A solution to this problem is a sequence of moves leading from the initial configuration to the goal configuration, and an optimal solution is a solution having the smallest number of moves. Not all problems have solutions; for example, in Fig. 1.1, Problem 1 has many solutions while Problem 2 has no solution at all.

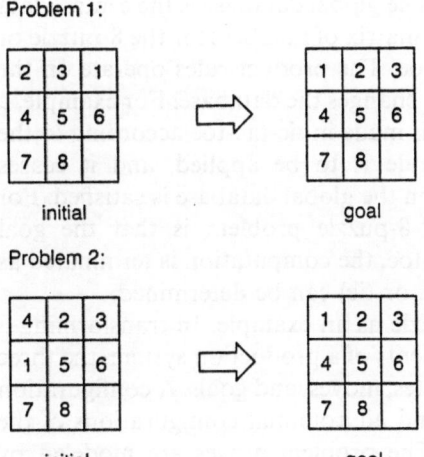

Fig. 1.1. Two 8-puzzle problems, the first one having many solutions and the second one having no solution

Tic-tac-toe is a two-person game in which players take turns at marking a 3×3 playing board. One player marks a cross ($\times$), and the other marks a circle ($\bigcirc$). Either player can play first, and whoever first marks a complete row, column, or diagonal wins. Figure 1.2 is a board configuration after both players have made their first moves. The main problem of an AI system, as one of the players in such a game, is to search for an optimal or a most promising choice before it makes each move.

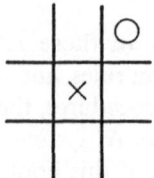

Fig. 1.2. A tic-tac-toe board configuration after both players have made their first moves

The essential difference between the 8-puzzle and tic-tac-toe is that the 8-puzzle is non-adversary, while tic-tac-toe is adversary. In an 8-puzzle problem, the move made at each step is tentative and may be ultimately discarded or modified for a final (optimal) solution. However, each move in tic-tac-toe is always permanent. After each move, the system playing the game waits for the opponent's response before making the next move, the opponent being another system or a person. The sequence of all moves made by both players forms a complete game and determines the end result of the game.

To solve a problem using an AI production system, we must first transform the problem statement into three components of a production system: the global database, the rules, and the control strategy. The global database is the central data structure of a production system, such as the matrix of numbers for the 8-puzzle or the playing board configuration for tic-tac-toe. The product rules operate on the global database, and the application of a rule changes the database. For example, a tile is slid in an 8-puzzle problem or a mark is made in tic-tac-toe according to the rules. The control system decides which rule is to be applied, and it ceases computation when a termination condition on the global database is satisfied. For example, the termination condition in an 8-puzzle problem is that the goal configuration has been achieved. For tic-tac-toe, the computation is terminated as soon as an end result of the game (win, loss, or tie) can be determined.

Consider a graph search using the 8-puzzle as an example. In transforming a graph-search problem into the three components of a production system, the three components often correspond to problem states, moves, and goals. A configuration of the global database is a problem state, and all potential configurations of the global database form the problem space. The problem moves are modeled by production rules that operate on the state descriptions. The problem states are connected by the production rules, and the problem space forms a (directed) graph. The problem goal is a goal state or a set of goal states, and a solution to a problem is a sequence of moves (i.e., a path in the graph) that transforms an initial state into a goal state.

In an optimal problem, a cost is assigned to each move, and the problem is to find a solution having minimal cost. For example, a constant cost, 1, may be assigned to each move in the 8-puzzle. Then an optimal solution is a path from the initial state to the goal state having the smallest number of moves, or the minimal cost.

One basic algorithm for solving a graph-search problem is that, starting from the initial state, the control strategy repeatedly chooses a search-tip node for expansion of the search tree until a goal state is produced. The system also keeps track of the rules that have been applied so that it can compose them into a problem solution. The control strategy consists in selecting rules and keeping track of both the sequences of rules already tried and the databases they produced.

In many problems, the domain of possible combinations of sequences from which to choose a solution is very large. An uninformed control strategy which selects an arbitrary node for expansion at each step soon generates a combinatorial explosion of possibilities that exhausts the capacities of even large computers. Many of these problems are NP-*complete problems*, for which the time taken by the best methods currently known grows exponentially with problem size.

In many AI systems, an essential characteristic of computations for selecting rules is the task-dependent information that these computations use. Knowledge about the problem domain is often the key to more efficient control strategies. Any information that helps reduce search cost (e.g., time or space) is called *heuristic information*, and a search procedure using such information is called a *heuristic search method*.

One important heuristic search method uses a so-called *evaluation function*, a function of the nodes. At a given node in a search tree, the evaluation function gives an estimate of the cost of a (minimal cost) path from the start node to a goal node constrained to go through this given node. This estimate is usually the sum of the minimal cost already found from the start node to the given node and an estimate of the minimal cost from the given node to a goal node. The control strategy uses an evaluation function to rank the search-tip node so that, at each step, a node with the least estimated cost is expanded.

For example, in the 8-puzzle, there are two common evaluation functions (Nilsson 1980, Pearl 1984) to estimate the cost from a given node to a goal node: the *number of misplaced tiles*, and the *city-block distance*. The number of misplaced tiles of a node is the number of tiles which are not matched between this node and the goal node. For example, the numbers of misplaced tiles of the initial states of Problem 1 and Problem 2 in Fig. 1.1 are 3 and 2, respectively. The city-block distance (also called the *Manhattan distance*) of a node is the sum of the horizontal and vertical distances of misplaced tiles between this node and the goal node. The city-block distances of the initial states of Problem 1 and Problem 2 in Fig. 1.1 are 4 and 2, respectively.

The selection of evaluation functions depends mostly on intuition, general rules, and experience. Two of many ideas that have been used to define evaluation functions are the probability that a node is on the best path and distance or difference metrics between a given node and the goal node (Pohl 1970).

The choice of evaluation function often critically determines search performance. For example, the city-block distance is often better than the number of misplaced tiles for 8-puzzle problems having solutions. Even so, both functions fail to detect the problems having no solution (e.g., Problem 2 in Fig. 1.1). In fact, each move of a tile corresponds to an even permutation of the 8 tiles. And a problem is solvable if and only if the permutation of the 8 tiles from the initial configuration to the goal configuration is even. Neither function reveals any information about the permutation of tiles. Therefore, the usage of heuristic information is crucial in AI graph-search techniques.

Different considerations exist in adversary games. At each step in playing an adversary game like tic-tac-toe, a system can, theoretically, generate a complete game tree (down to the termination of the game) and then search for an optimal move by backing up (e.g., using the minimaxing procedure) the end results at the terminations. However, such a search into the terminations of a game is feasible only for very simple games. For most games, such as complete chess and checker games, only a limited portion of a complete game tree can be actually generated because of combinatorial explosion. Therefore, most AI game-playing systems, like general graph-search problems, base their decision making on heuristic information.

A heuristic search method usually generates only a portion of a game tree, using a *static evaluation function* to evaluate the search-tip nodes and then determining a move by backing up these evaluated estimates at the search-tip nodes. Since the heuristic information involved in the evaluation of the search-tip nodes is usually not perfect, a move made by a heuristic method is generally not optimal. The decision making at each step depends on both the search-depth and the usage of heuristic information. The use of heuristic information may consist of, for example, use of an evaluation function and a back-up process. Since each move (by either player) is permanent once made and cannot be modified afterwards, the quality of performance of a game-playing system using heuristic methods depends on both the decision making of the system at each step and the opponent's responses.

Since heuristic decision making at a move depends on the search game tree which is only a small portion of the complete game tree, it is natural to consider the relationship between the search tree and decision making. It is intuitively believed that the larger the search tree is, the better the decision making is. And computer programs playing common games such as chess and checkers really improve their performances with increasing search-depth. However, in some game models, a recently found (pathological) phenomenon indicates that decision making sometimes deteriorates when the size of the search tree becomes larger (Beal 1980, Nau 1980, Pearl 1983).

Since the use of heuristic information is essential in many AI techniques, such as graph searches and game-tree searches, the study of heuristic information is important in AI. Such a study should include the properties of heuristic information, especially the properties relevant to solution of each particular problem.

How to search for useful heuristic information and how to use heuristic information are also important.

In the following chapters, we introduce a theoretical study of heuristic information using probabilistic games as the problem domains. In this study, heuristic information is precisely formulated, and a mathematical theory of decision making based on heuristic information is developed. The goal of this study is to build a foundation for the use of heuristic information. The contents of each chapter are outlined below.

In Chap. 2, basic notions of finite, zero-sum, two-person, perfect information games, including minimax values, are introduced.

In Chap. 3, conventional heuristic game-tree searches consisting of static evaluations and minimaxing processes are introduced. The recently found pathological phenomenon and the newly proposed back-up processes are also discussed.

Chapter 4 briefly presents the concepts of probability theory needed in our study. The idea of martingale used for representing a system of estimation is also included.

In Chap. 5, probabilistic game models which will be used as the problem domains in our study, P_b- and G_d-game models being two typical examples, are formulated. The concept of game values relative to players' strategies is also introduced. Game values, with minimax values as special cases, are used for representing the strength of nodes in game trees.

In Chap. 6, heuristic information and its accuracy are specifically formulated. A formal definition of heuristic search as a process for collecting heuristic information at each node of a game tree is presented. A basic property of a heuristic search is that it collects more precise heuristic information if it searches more nodes.

Chapter 7 presents a formal definition of any type of strength of a node as a random variable. Based on heuristic information, estimation of the strength of a node as the estimation of a random variable is then explained. That the estimate of a random variable becomes more precise if it depends on more precise heuristic information is one important result presented here. A decision model and a decision strategy that improves decision quality and does not show the pathological phenomenon are then formulated for our abstract model.

In Chap. 8, the fact that a back-up process using the product-propagation rules should be used for estimating minimax values in independent, WIN-LOSS game models is demonstrated.

Chapter 9 discusses the estimation of minimax values in P_b-game models. The estimate, completely derived at a search-tip node, and based on a heuristic search using a specific evaluation function called the one-counter, is presented. The back-up process in this estimation, using the product-propagation rules, is also discussed.

In Chap. 10, G_d-game models are considered. The estimate of the minimax value of a node is completely derived. The heuristic search, using the one-counter as an evaluation function and requiring no back-up process in the estimation of minimax values, is presented. That the actual process of estimating node strength may be very different for different games is demonstrated by examples in Chaps. 9 and 10.

In this book, heuristic information involved in non-probabilistic games (e.g., chess and checker games) is not discussed, and no new actual search process for general games is proposed. The examples in this book are simple and mainly of theoretical interest. However, starting with probabilistic game models, this book is an attempt to study theoretical backgrounds of heuristic information in AI and to build a solid foundation for further exploration. Only with an accurate theoretical base can development in this field be sound and of optimal use in meeting the need for more accurate heuristic information in the AI field.

2 Games and Minimax Values

This chapter introduces some basic concepts of games. Finite zero-sum two-person perfect information games, for which our theory is to be developed, are introduced first. These games are represented by special graphs, called game graphs. If nodes representing the same game position in a conventional game tree are merged into only one node, then the merged graph is called a game graph. Two examples are used to clarify the discussion.

Minimax value, an important notion in conventional game theory, and the corresponding minimax and alpha-beta procedures are also introduced in this chapter. The games and game-tree searches discussed here are theoretical and idealized. That is, both players are searching the tree out to the leaves. The heuristic searches discussed later in the book are based on this idealized consideration.

2.1 Finite Perfect Information Games and Game Trees

Games are played, in general, by $n \, (> 0)$ persons. Suppose that all games terminate after a finite number of moves and that at any stage of play there are only a finite number of choices for the next move. Such games are called *finite games*.

A move decided only by free decision of a player is called a *personal move*. This personal move can be characterized by the set of all alternatives of the move $\{A_1, \ldots, A_k\}$ $(k \geq 1)$ associated with the player. A chance move is a move decided by some mechanical device which makes a choice with definite probabilities, similar to shuffling a deck of cards or throwing a die. A chance move can be characterized by the set of all alternatives $\{A_1, \ldots, A_k\}$ $(k \geq 1)$ associated with their probabilities $\{p_1, \ldots, p_k\}$, where $p_i \, (\geq 0)$ is the probability of A_i for each i, and $\Sigma_i \, p_i = 1$.

Consider a special category of games, called *perfect information games*. In a game with perfect information, each player is completely informed at each move about all previous moves, personal or chance. That is, the decision made at each of the previous moves is completely known to each player before the next move. For example, tic-tac-toe and chess are perfect information games without chance moves; backgammon is a perfect information game with chance moves. Bridge and poker are not perfect information games since each player does not know the cards dealt to other players.

Consider any finite, perfect information game and imagine such a game represented by a tree of nodes. The root of the tree is the initial state from which it is the first player's turn to move. The successors (or sons) of the root are all possible states the first player can reach in one move. If the first player has n alternatives, then the root has n successors, each representing one of the alternatives. If the first move is a chance move, each link of the root, pointing to a successor, is also associated with the probability of this successor. From each of the successors of the root, it is again the turn of a player to move, and all these alternatives form the sons of this node, and so on. The tree is built in this way until play is terminated.

Each non-terminal node of the tree is characterized by the player who is to move from the node and by a set of links, each link representing one possible move and pointing to the next corresponding state of play. In the case of a chance move, each link is also associated with the probability of the corresponding alternative. A terminal node is a state at which play ends; this node is characterized by the resulting payoff to each player.

2.2   Zero-Sum Two-Person Perfect Information Games

In this section we consider perfect information games with two players, who move alternately. There are no chance moves. Such games are here called two-person perfect information games, and the two players are called LEFT and RIGHT.

If we assume that the players make payments to each other at the end of each game, then at each terminal node the sum of the payoffs of both players is zero. That is, if the payoff to LEFT at a terminal node is A, then the payoff to RIGHT at that same terminal node is $-A$. Such games are called *zero-sum games*.

In playing zero-sum games, suppose that the goal of the players is to maximize their own payoff when the game ends. (This is only an idealized assumption. The actual goal of playing may be different. Maximizing expected payoff, maximizing minimum payoff, and other objectives are all reasonable goals. See Sect. 7.3 for a precise formulation). Since both players make payments to each other at the end of the game and both payoffs at each terminal node have sum zero, we assign to each terminal node only the payoff of a fixed player. The goal of this fixed player is then to maximize these values (her payoffs) assigned at the terminal nodes. For the other player, if the value assigned at a terminal node is A, then his payoff at that terminal node becomes $-A$. Therefore, he should minimize the values at the terminal nodes in order to maximize his own payoffs.

In summary, we are considering games with the following properties:

1. There are only two players, who move alternately.
2. From any possible position there are a finite number of choices for the next move, and the game terminates after a finite number of moves.
3. Each game is represented by a finite tree, called the corresponding *game tree*. Each non-terminal node is a position from which it is a player's turn to move,

and its successors are all of the positions which it is possible to reach in one move. Further, each terminal node is associated with a value, called a *payoff.*

4. Both players know completely about all moves previous to a particular move; that is, they have *perfect information.*
5. The goal of the players is that one wishes to maximize the payoffs at terminal nodes and the other wishes to minimize them.

We call all games with the above 5 properties *zero-sum two-person perfect information games.* The player who wishes to maximize the values at terminal nodes is called *MAX* and the other player *MIN.* In a game tree, nodes from which it is MAX's turn to move are called *MAX nodes,* and other nodes are called *MIN nodes.*

2.3 Subgames and Game Graphs

After each move of a zero-sum two-person perfect information game, the new game position can be treated as the initial position of another game, having the same rules and being a *subgame* of the original game. Therefore, a game can be seen as a *hierarchy of subgames,* and each position of the game starts a new subgame. In the corresponding game tree, the subtree under each node is the game tree of the corresponding subgame.

In a game tree, different nodes may represent the same game position. The subgames from all of the nodes representing the same game position are identical. If we merge all of the nodes representing the same game position to only one node, the game tree becomes a directed graph, in which each (directed) edge represents a move from one position to another. A terminal node in the game tree is also a terminal node (without successors) in this merged graph, representing an end of the game. The root in the game tree is the only node without predecessors in the merged graph. Moreover, since we are considering finite games, there are no cycles in such a directed graph, and any possible path starting from the root always ends at a terminal node, in only a finite number of edges. We call such a graph a *game graph.*

In game-playing programs, games are usually represented by a game tree; however, at times the corresponding game graph is more easily analyzed than the game tree. In a game graph, the interrelation between nodes is explicitly represented. Nevertheless, we will not emphasize the difference between game trees and game graphs. In fact, a game tree is a special game graph. All discussions in this book are suitable for general game graphs.

In a game graph, the *height* of a node is defined as the number of edges in the longest path from that node to a terminal node. If a node N is a terminal node, then its height $\text{HEIGHT}(N) = 0$. If N is a non-terminal node and has $N_1, \ldots, N_k$ as its successors, then its height satisfies:

$$\text{HEIGHT}(N) = 1 + \max \{\text{HEIGHT}(N_1), \ldots, \text{HEIGHT}(N_K)\}.$$

We also use the term *level* to denote a set of all nodes in a game tree which can be reached from the root in the same number of edges. The root is the first level of the tree; the second level is the set of all successors of the root; their successors form the third level, and so on. The nodes at the same level may be of different height. Since there may be different paths from the root to a node in a game graph, the notion of level, defined for a game tree, is not suitable for a general game graph.

2.4   Example 1: G_1-Games

The first game to be used as an example consists of a row of a fixed number of squares each of which is covered by a white or black chip. The color of the chip on each square is chosen independently by a fixed random machine. The two players, called WHITE and BLACK, take turns at removing one chip from either end of the row. WHITE wins if the last remaining chip is white, and BLACK wins if it is black. We call such a game a G_1-game. It is a special type of G_d-game, in which each player removes d ($d \geq 1$) chips at each step (Chap. 5). For a G_d-game, the number of chips should be $nd + 1$, where n is any non-negative integer. Let the integer 1 represent a white chip and 0 a black one. Then WHITE becomes MAX, who wishes to maximize the value at the end of the game, and BLACK becomes MIN, who wishes to minimize the value.

Figure 2.1 is the game tree of a G_1-game, where each node is represented by the row of 1's and 0's corresponding to the board configuration. The left son of a node is the board configuration obtained by removing a chip from the right end, and the right son is that obtained by removing a chip from the left end. Note that, no matter

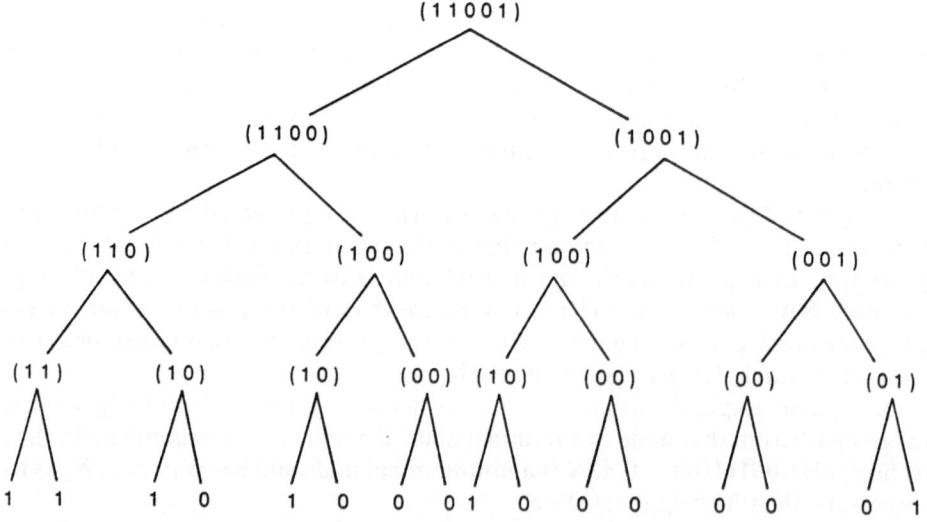

Fig. 2.1. A G_1-game tree

what kinds of chips are on the board, removing a chip from the right end followed by removing a chip from the left end has the same result as removing the leftmost chip followed by removing the rightmost chip. Both cases produce the same board configuration. Therefore, we can use a simpler game graph to represent a G_1-game. Figure 2.2 is the game graph corresponding to the G_1-game in Fig. 2.1. For this special game graph, the level of nodes can be similarly defined as for game trees.

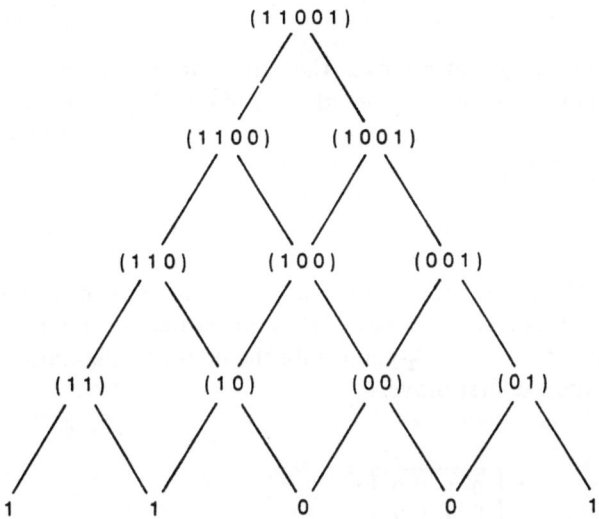

Fig. 2.2. The game graph corresponding to the G_1-game in Fig. 2.1

In each level of such a game graph, any two consecutive nodes have a common successor. We call such a graph a G_1-game graph and use it to analyze G_1-games. Therefore, a G_1-game is characterized by its G_1-game graph, in which terminals are independently assigned the value 1 or 0 with the probability p or $1-p$ ($0 \leq p \leq 1$), respectively. G_1-games were introduced by Nau (Nau 1983b) in a study of the pathology in heuristic game-tree searches.

2.5 Example 2: P_2-Games

The next example is a board-splitting game. Each game consists of an $N \times N$ board (N is a power of 2). The initial board configuration is set up by using a device to cover random cells of the board by black chips, so that each cell is chosen independently with the probability p ($0 \leq p \leq 1$). Starting with such a board configuration (e.g., Fig. 2.3), two players, called VERTICAL and HORIZONTAL, take turns at splitting the board into two halves and removing the unwanted half. VERTICAL splits the board vertically and removes the right or the left half;

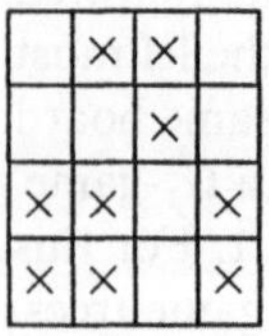

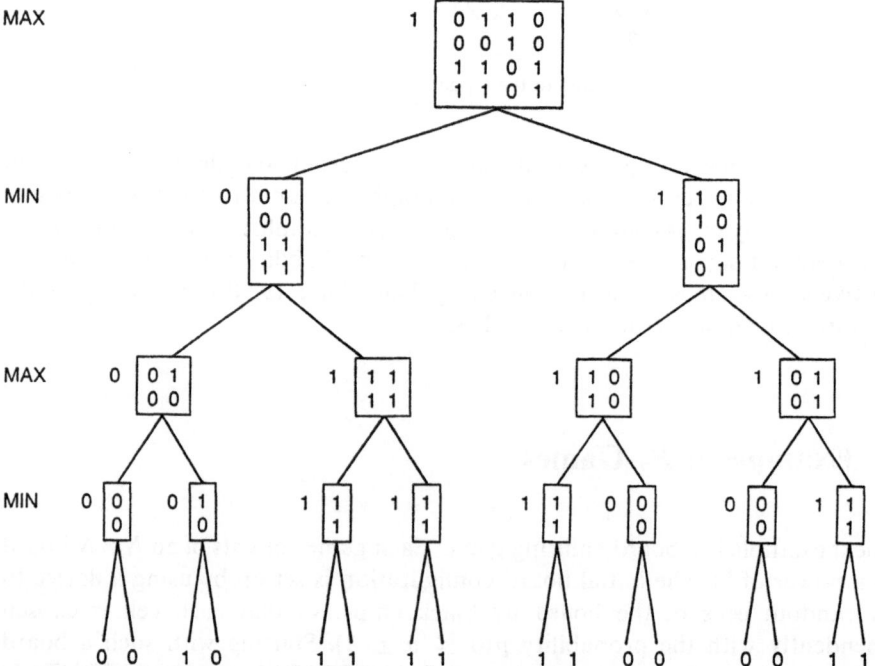

Fig. 2.3. A playing board of a P_2-game, where each grid covered by a chip is marked by an "×"

HORIZONTAL splits it horizontally and removes the upper or the lower half. VERTICAL wins if the last remaining cell is covered by a black chip, and loses otherwise. We call such a game a P_2-game. It is a special type of P_b-game, in which the board is split into $b(b \geq 2)$ parts at each move (Chap. 5).

A P_2-game can be represented by a complete binary tree, in which a node shows the corresponding board configuration. A terminal node has only one cell. Assign the value 1 to all of the cells covered by black chips and the value 0 to all of the uncovered cells. Then VERTICAL becomes MAX and wishes to maximize the values in the cells, and HORIZONTAL becomes MIN and wishes to minimize the values. Figure 2.4 is the game tree of the P_2-game with the board configuration in Fig. 2.3, and MAX is to make the first move.

Fig. 2.4. The game tree of the P_2-game in Fig. 2.3, where the number at each node is the minimax value of the node

At each level of the game tree of a P_2-game, all of the board configuration have the same size and are completely independent. Therefore, the corresponding game graph is the game tree itself. In fact, a P_2-game is characterized by a finite complete binary tree, in which all terminal nodes are independently assigned 1 or 0 with a probability p or $1 - p$ $(0 \leq p \leq 1)$, respectively. P_2-games are discussed in detail by Pearl (1980, 1983) and Nau (1983b).

2.6   Minimax Values, the Minimax Procedure and the Alpha-Beta Procedure

Consider a zero-sum two-person perfect information game, which is represented by a game graph with payoffs assigned to terminal nodes. One player, MAX, wishes to maximize the payoffs, and the other player, MIN, wishes to minimize them. For the analysis of such a game we introduce the notion of the minimax value of a node in the game graph. This minimax value depends only on the game graph itself and represents a compromised payoff if both players play the game from this node in an optimal way.

The minimax value of each node is defined in the following recursive way; its existence will become clear in the process of the assignment.

1. The minimax value of a leaf (i.e., a node of height 0) is the payoff already assigned to the leaf. A leaf represents an end of the game, and the value at the leaf is the result (i.e., the payoff).
2. Suppose that N is a node of height $k(> 0)$ and that the minimax values of nodes of height less than k have been defined; then the minimax value of N is the largest (or smallest) of the minimax values of its successors if N is a MAX (or MIN) node.

In the above definition, since the successors of the node N are of height less than k and their minimax values are assumed to be already defined, the minimax value of N is thus well defined. In game theory the minimax value of the root of a game graph is also called the *game value* of the game (Dresher 1981, von Neumann and Morgenstern 1946); therefore, the minimax value of a node N is the game value of the subgame starting from the node N.

The game value of a game plays a central role in game theory and, if both players play optimally, is a compromised result of that particular round. This can be obviously seen if the game tree consists of only one leaf. Now suppose that the root of the game tree has successors and that the minimax value of each successor is known and represents the compromised result (payoff) of the subgame starting from the successor. First let the root be a MAX node. MAX, wishing to maximize the payoffs, should choose a successor with the maximum minimax value, and thus this maximum minimax value becomes the game value. If the root is a MIN node, MIN should minimize the payoff by choosing a successor with the minimum minimax value, so that the minimum becomes the game value.

Now let the minimax value of any node N be v. Suppose that MAX and MIN play the game starting from the position N. The value v means that, no matter how MIN plays, MAX can possibly get a payoff of at least v. In order to achieve this goal, MAX needs only to choose a successor with the highest minimax value among all successors at each MAX node. On the other hand, no matter how MAX plays, MIN, who wishes to minimize the value, can possibly get a payoff of at most v. In order to achieve this goal, from each MIN node, MIN needs only to choose a successor with the smallest minimax value. Such a move by MAX or MIN is called a *minimax optimal move*. MAX (or MIN) is said to play optimally at a position if a minimax optimal move is made. If both MAX and MIN play optimally, the game results in the exact payoff v.

Although it is easy to demonstrate that choosing a minimax optimal move is not necessarily the best way to play if the opponent does not play optimally, the discussion here focuses on optimal strategies. In order to play optimally from a node, MAX should find a successor with the largest minimax value, and MIN should find a successor with the smallest minimax value. Therefore, the determination of an optimal move depends completely on the minimax values of possible next moves and, thus, finding minimax values is vital in determining such optimal game playing.

Note that any procedure for finding minimax values needs to generate the tree down to the leaves. Therefore, optimal play requires both players to search to the leaves. Optimal play is idealized play and is feasible only for simple games. The following are two known procedures for finding minimax values.

A. Minimaxing

Consider the minimax value $M(P)$ of a game position P. The most straightforward way of finding $M(P)$ is to generate the corresponding game graph (or game tree) down to the leaves, with the game position P as the root, and then to back up the actual payoffs at leaves using the so-called minimaxing operations, which are summarized as the following recursive procedure.

The Minimax Procedure $M(P)$

1. If P is terminal, return $M(P) = $ the actual payoff at P. Otherwise, let $P_1, P_2, \ldots,$ P_k be the successors of P; if P is a MAX node, go to Step 2, else go to Step 3.
2. Set $M(P) = \text{Max}_i\ M(P_i)$.
3. Set $M(P) = \text{Min}_i\ M(P_i)$.

The minimax procedure generates the complete game tree and backs up the values at terminals so that the minimax value of each node is returned. An example is given in Fig. 2.5.

Note that the above procedure is designed according to the definition of minimax values and that it involves a complete generation of the game graph and an evaluation of every node in the game graph. Since the goal is to find the minimax value of just the root, the minimax procedure is usually an inefficient strategy.

MAX

MIN

MAX

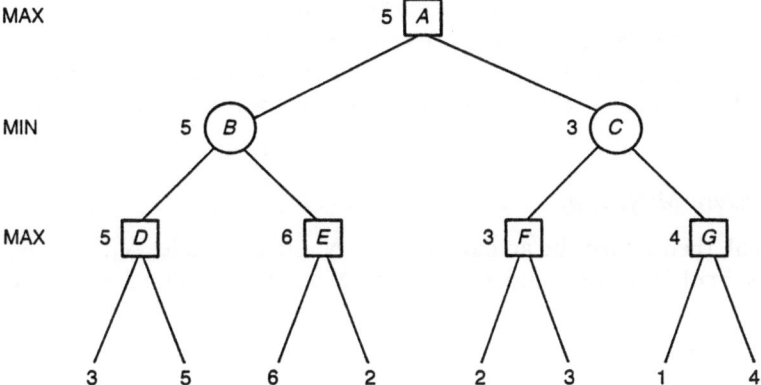

Fig. 2.5. A minimaxing process

B. Alpha-Beta Pruning

The alpha-beta pruning pocedure is the most celebrated procedure used to speed up a game-tree search for the minimax value of the root. To obtain the minimax value of a node, it is not always necessary to search for the exact minimax value of each descendant of the node. First, note the following facts relevant to minimax values.

Fact 1. For a MAX node, if it is known that the minimax value of one of its successors (MIN nodes) is a, and it is also certain that all minimax vales of other successors are at most a (exact values are not necessary), then the minimax value of this MAX node is a. Therefore, given the minimax value a of a successor, the search for the minimax value of another successor can be terminated as soon as it can be determined that the search value is at most a. Such a value a is called a lower cutoff (or an alpha cutoff)..

Fact 2. If a successor (a MAX node) of a MIN node has at most a as its minimax value, then this MIN node has at most a as the minimax value no matter what values other successors have.

Fact 3. For a MIN node, if the minimax value of one of its successors (MAX nodes) is b, and if it is also certain that all minimax values of other successors are at least b (exact values are not necessary), then the minimax value of this MIN node is b. Therefore, given the minimax value b of a successor, the search for the minimax value of another successor can be terminated as soon as it can be determined that the search value is at least b. Such a value b is called an upper cutoff (or a beta cutoff).

Fact 4. If a successor (a MIN node) of a MAX node has at least b as its minimax value, then this MAX node has at least b as the minimax value no matter what values other successors have.

The above considerations can be developed into the following recursive procedure which we call $M(N; \alpha, \beta)$. This procedure, with two parameters, $\alpha \leq \beta$, evaluates $M(N)$, the minimax value of N, if $M(N)$ lies between α and β. Otherwise, it returns either a value $\leq \alpha$ (if $M(N) \leq \alpha$) or a value $\geq \beta$ (if $M(N) \geq \beta$). Therefore, the minimax value of the root N can be obtained by the procedure called $M(N; -\infty, \infty)$.

Alpha-Beta Procedure $M(N; \alpha, \beta)$

1. If N is terminal, then return the actual payoff at N. Otherwise, let $N_1, \ldots, N_k$ be the successors of N; set $i = 1$ and, if N is a MAX node, go to step 2, else go to step 2'.
2. Set $\alpha = \max \{\alpha, M(N_i; \alpha, \beta)\}$.
3. If $\alpha \geq \beta$, return, β; else continue.
4. If $i = k$, return α; else set $i = i + 1$, and go to step 2.
2'. Set $\beta = \min \{\beta, M(N_i; \alpha, \beta)\}$.
3'. If $\beta \leq \alpha$, return α; else continue.
4'. If $i = k$, return β; else set $i = i + 1$, and go to step 2'.

In the example of Fig. 2.6, the procedure $M(A; -\infty, \infty)$ for evaluating the minimax value of the node A calls the procedure $M(B; -\infty, \infty)$, which itself calls $M(D; -\infty, \infty)$. The procedure $M(D; -\infty, \infty)$ returns the actual minimax value 5 of the node D; this value 5 becomes an upper cutoff for its sibling, and the next procedure call is hence $M(E; -\infty, 5)$, which only searches for the value 6 of the left successor and leaves the right successor not searched. Therefore, the procedure $M(B; -\infty, \infty)$ returns the value 5 which becomes a lower cutoff for the subtree rooting at the node C; that is, the next procedure to be called is $M(C; 5, \infty)$, which returns 5 and leaves its right subtree not searched since $M(F; 5, \infty)$ already

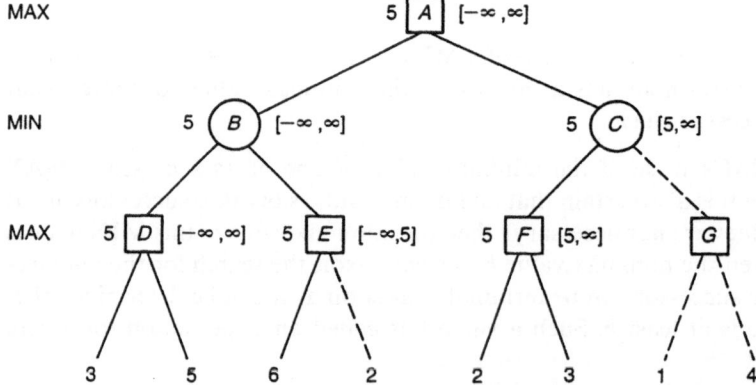

Fig. 2.6. An alpha-beta pruning process, in which there is an alpha-beta procedure call at each searched node. The pair at the right of the node shows the parameters of the procedure call, and the value at the left is the value returned by the call

determines that the minimax value of F is at most 5. This process leaves one internal node and three terminal nodes unsearched. Because it searches fewer nodes, in this example the alpha-beta procedure is more efficient than the minimax procedure.

Knuth and Moore (1975) have used a "negmax" approach to prove the correctness of the alpha-beta procedure and have shown that the alpha-beta procedure is optimal in a certain reasonable sense. Pearl (1984) and Baudet (1978) have given a detailed analysis of this procedure.

3   Heuristic Game-Tree Searches

In the last chapter, we discussed zero-sum two-person perfect information games. If both players playing such a game can find at each step the minimax values of all possible next moves, then the corresponding game-playing problem is completely solved. MAX chooses a move of the highest minimax value at each step, and MIN chooses a move of the lowest minimax value at each step. The moves chosen have identical minimax values; this value is exactly the same as the final payoff of the game. It is because this use of minimax values solves game problems that finding minimax values is a central issue in the theory of zero-sum two-person perfect information games.

However, in finding the actual minimax value of a node in a game tree, the number of terminal nodes inspected by any algorithm is usually exponential, relative to the height of the node (Pearl 1983). For example, a complete game tree for the game of checkers has been estimated as having about 10^{40} non-terminal nodes (Samuel 1959). Even if 3 billion nodes could be generated each second, generating the whole tree would still require around 10^{21} centuries. Except for very simple cases, therefore, the above theoretical game-playing strategy using actual minimax values is simply not feasible.

In this chapter, we first introduce the conventional strategy applied in game-playing programs. Since it is not feasible to search to the leaves, a game-playing program usually generates only a reasonable portion of the game tree and evaluates the search-tip nodes using a static evaluation function. The values returned by the static evaluation function are treated as if they were true terminal payoffs and are backed up by the minimaxing process as in the idealized cases of the last chapter. A move is then decided by choosing a node with the most promising backed-up value. Many successful game-playing programs for playing common games such as chess and checkers use this heuristic search strategy.

Minimaxing is only a heuristic for the choice of move. There is no theoretical justification. Although computer programs for playing common games can improve their performance with increasing search-depth, a pathological phenomenon, discussed in this chapter, was found in the study of game-tree searches. That is, the performance of the decision in this conventional method deteriorates with increasing search-depth for some special games. Also, new back-up procedures have been demonstrated to be sometimes better than minimaxing. These procedures are introduced in this chapter.

This unrest was the original motivation for developing the mathematical model of heuristic game-tree search introduced later in this book.

Without loss of generality, all games are considered from MAX's point of view.

3.1   The Conventional Heuristic Game-Tree Search

Since there are usually no practical ways of finding the actual minimax value of a node, one may naturally use a heuristic approximation. Conventionally, only a reasonable portion of a game tree rooted at the given node is generated. Then a "static evaluation function" is used to estimate the potential strength of all frontier nodes, and those estimates are backed up to represent the potential strength of the ancestor nodes up to the given node. A move (MAX's) is then decided by choosing the node with the highest estimate, as if those estimates were actual minimax values. Many successful game-playing programs have been developed by using this method.

3.1.1   Static Evaluation Functions

There are two essential components to the conventional heuristic method: the static evaluation function used for estimating search-tip nodes, and the back-up process used for estimating ancestor nodes.

A static evaluation function believed to estimate the potential strength of a node is usually chosen by a heuristic approach and depends on the individual game. For example, experience shows that certain features in a game position contribute to the strength of the position (for MAX) whereas others tends to weaken it. The value returned by a static evaluation function at a node is called a static value at the node (relative to the static evaluation function).

Consider the game tic-tac-toe, for instance. Suppose that the first player, MAX, marks a cross ($\times$), and the second player, MIN, marks a circle ($\bigcirc$). For a position p, $e(p)$ is the value defined as follows (Nilsson 1980).

If p is not a winning position for either player, $e(p) =$ (number of complete rows, columns, or diagonals that are still open for MAX) $-$ (number of complete rows, columns, or diagonals that are still open for MIN).

If p is a win for MAX, $e(p) = \infty$ (denotes a very large positive number).

If p is a win for MIN, $e(p) = -\infty$.

Thus, if p is the position in Fig. 3.1, then $e(p) = 6 - 4 = 2$. Since the player who first marks a complete row, column, or diagonal wins, the number of such arrays still open for one player indicates a certain type of chance for that player to complete such an array. Therefore, the difference between these two numbers indicates a certain type of chance for MAX to win and can be used as a static evaluation function. Of course, the above argument rests merely on common sense. An exploration of the actual effectiveness of this function demands rigorous analysis.

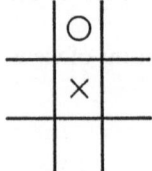

Fig. 3.1. A tic-tac-toe game position.

Next, consider G_1- and P_2-games (introduced in Chap. 2). A position can be represented as an array of 0's and 1's, one-dimensional for a G_1-game and two-dimensional for a P_2-game. MAX wins such games if the last remaining entry is 1. Intuition suggests that a position with more 1's is more favorable to MAX. Thus, a count of the number of 1's in a non-terminal position N can be used as a static evaluation function to indicate the strength of a position for MAX (Nau 1983b):

$$L(N) = \text{the number of 1's in } N.$$

Since the total number of entries is the same for all of the nodes at the same level in the game tree, the number of 0's in a position need not be considered separately.

The one-counter as a static evaluation function for general G_d- and P_b-games will be studied in detail later in this book.

3.1.2   The Back-Up Process

The second part of the heuristic method is the back-up process, which is used to estimate the ancestor nodes by backing up the values returned by a static evaluation function at the search-frontier nodes. The conventional back-up method is based on the face-value principle (Pearl 1984). In the face-value principle, the values returned by the static evaluation function are treated as if they were the actual minimax values of the search-frontier nodes. Then, relative to these values, the returned minimax values of the ancestor nodes, derived for example by the minimax procedure or the alpha-beta procedure (simply a method to speed up the minimaxing process), are treated as estimates of the ancestor nodes. This face-value principle results in the conventional minimax procedure and the alpha-beta procedure's being the most popular back-up processes in this heuristic approach.

Consider the P_2-game in Fig. 2.3 for example. Figure 3.2 is a partial game tree of that game, where MAX is to move from the root. The one-counter has the value 5 at the left successor of the root and the value 4 at the right successor. Considering these two values, MAX would choose a move to the left successor. However, relative to the values of the one-counter at the tip nodes in Fig. 3.2, the left successor of the root has the backed-up value 1, and the right successor has the backed-up value 2. Therefore, MAX would move to the right successor on the basis of these two backed-up values. Figure 2.4 shows that the actual minimax value of the left successor of the root is 0 and the actual minimax value of the right successor is 1. It is thus clear that, in this case, backed-up values are superior to the values

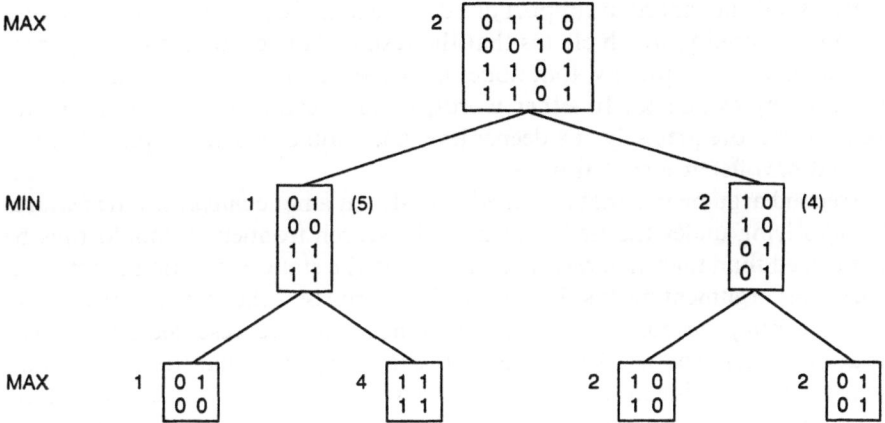

Fig. 3.2 A heuristic search game tree of the P_2-game in Fig. 2.3, using the one-counter as a static evaluation function.

returned directly by the one-counter at both successors of the root because they bring MAX closer to the actual minimax value.

3.2   Heuristic Arguments and the Pathological Phenomenon

The use of the heuristic search method introduced in the last section is primarily based on experience. Many successful game-playing programs for playing common games such as chess and checkers use this search method. Most programs search the game tree to a fixed depth. There is a varied-depth search called "iterative deepening," which works as follows. A complete search is done to depth 2, returning a move. The search is then redone to depth 3, again to depth 4, and so on, until a preset time limit is exceeded. A program using this iterative-deepening search has already achieved an expert rating in human play (Barr and Feigenbaum 1981).

However, since the static evaluation function is only an approximation and may be wrong, minimaxing no longer serves its original purpose of defining and identifying a theoretically correct move. Instead, minimaxing is only a heuristic for the choice of move. In the study of heuristic game-tree search, many different back-up procedures and the performance as a function of search depth have been investigated in special games.

First, consider search depth in the conventional heuristic search. It is usually presumed that the backed-up evaluation gives a more accurate estimate than would be obtained by applying the static evaluation function directly to those moves. It is also presumed that a deeper search gives more accurate backed-up values, and, therefore, that a deeper search would improve the performance of the

decision, as in computer programs playing common games. Two heuristic arguments are usually advanced in support of a deep search (Pearl 1984). The first one is
the notion of visibility, which claims that the result of the game is more apparent
near its end and, consequently, that nodes at deeper levels of the game tree can be
more accurately evaluated. In other words, a static evaluation function should
estimate nodes more precisely at a deeper level and should return complete features
(e.g., actual payoffs) at terminal nodes.

The second argument is that a backed-up value at a node *integrates* the features
of all nodes lying under the node down to the search frontier. It should thus be
more informed than the value returned directly by the static evaluation function at
the node. This argument means that the back-up process – the minimax procedure
or the alpha-beta procedure – integrates all of the features represented by the static
values at the search frontier. Therefore, it is usually expected that the performance
of a heuristic method like that described above will be improved by a deeper search.

For a more precise formulation of this idea, the *performance quality* of a
heuristic process is defined as the actual minimax value of the node chosen by the
process. Since actual minimax values reflect the strength of the nodes, it is expected
that this performance quality would be improved in a deeper search, that is, on the
average, the node chosen is one with a higher actual minimax value as the search
deepens.

However, recent studies (Beal 1980, Nau 1980, Pearl 1983) found a *pathological*
phenomenon in the conventional heuristic method: the performance quality may
decrease in a deeper search. In other words, the node chosen is more likely, as the
the search becomes deeper, to be one with a lower actual minimax value in many
games. Consider P_b-games for example. With the one-counter as the static evaluation, a deeper search chooses fewer nodes of forced win (minimax value = 1) than a
shallower one (Nau 1983b).

Therefore, for choosing a node of forced win, it is better to use the one-counter
directly for decision making than to search several levels by the minimaxing
process.

In other words, if we assume that either a static evaluation function itself or a
back-up process (e.g., the minimax procedure or the alpha-beta procedure) returns
estimates of the actual minimax values, then the pathology indicates that the values
returned directly by the static evaluation function are better estimates than the
backed-up values for some cases. Therefore, the so-called face-value principle fails
sometimes in estimating actual minimax values. In other words, the minimaxing
process is not estimating minimax values.

For detailed discussion of this pathological phenomenon, the reader may refer
to the references mentioned above.

3.3   New Back-Up Processes

Since the conventional method in a heuristic game-tree search has been demonstrated to be suboptimal for some games, many studies on new methods have

been conducted. These studies are primarily experimental and have concentrated on the back-up process. This section introduces some of these newly studied back-up processes.

3.3.1   The Product-Propagation Procedure

Consider games with only two possible results: WIN and LOSS (from MAX's point of view), as in the cases of G_1- and P_2-games. Suppose that a static evaluation function at a searched node gives the (conditional) probability of being a WIN node (or a forced WIN), relative to certain search features. Here a WIN node means that the actual minimax value of the node is a WIN (which is 1 in G_1- and P_2-games) and that from this node MAX can force a WIN at the end of the game.

Let $p_1, p_2, \ldots, p_n$ be the values returned by the static evaluation function at the sons of a node A (Fig. 3.3). These p_i's are probabilities of each son's being a WIN node.

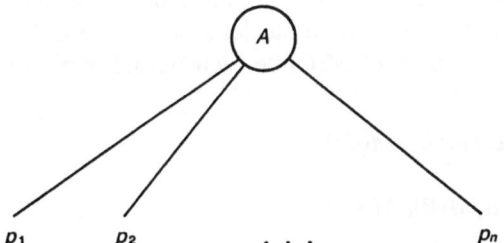

Fig. 3.3. Values returned by a static evaluation function at the sons of a node A.

Let p be the conditional probability that A is a WIN node, given those p_i's. There are two different cases to consider.

Case 1: A is a MAX node. A is a WIN node if and only if one of its sons is a WIN node, or A is a LOSS (i.e., not a WIN) node if and only if all of its sons are LOSS nodes. Therefore, the conditional probability that A is a LOSS node is

$$1-p=\prod_i (1-p_i)$$

if all of its sons are independent. The probability p of a WIN at A is

(I) $$p=1-\prod_i (1-p_i).$$

Case 2: A is a MIN node. Then A is a WIN node if and only if all of its sons are WIN nodes. Therefore, the probability p of A being a WIN node is

(II) $$p=\prod_i p_i$$

if all of its sons are independent.

The rules (I) and (II) are called the *product-propagation rules*, and the corresponding back-up process is called the *product-propagation procedure*, first proposed by Pearl (1981). This procedure supposes that the initial values at the search frontier should be between 0 and 1 (inclusive) and represent the probabilities of a WIN node. If a static evaluation such as the one-counter in G_1- and P_2-games does not return values of this type, these values should be transformed to the appropriate form before the product-propagation procedure is applied.

In Chaps. 9 and 10, the one-counter is studied in detail for both P_b-games and G_d-games, and the formulas which transform the values of the one-counter into conditional probabilities are completely derived.

3.3.2 The *M* & *N* Procedure

Before the discovery of the pathological phenomenon, Slagle and Dixon (1970) had already noticed that the minimaxing process is not always an optimal way to estimate minimax values and had proposed the so-called *M* & *N* procedure. In their discussion, a static evaluation function is assumed to return at a node an estimate of the mean (or the average) of a distribution. For example, let a node A (Fig. 3.4) have two sons B and C, and let $M(A)$, $M(B)$, and $M(C)$ be their actual minimax values. Then

$$M(A) = \max\{M(B),\ M(C)\}$$

or

$$M(A) = \min\{M(B),\ M(C)\}$$

if A is a MAX or a MIN node, respectively.

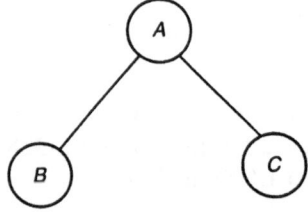

Fig. 3.4. A node, A, with two sons, B and C

Suppose that V_1 and V_2, either backed-up values or values returned directly by the static evaluation function, are estimates of the means of $M(B)$ and $M(C)$:

$$V_1 \approx m_1 = \text{the mean of } M(B),$$

$$V_2 \approx m_2 = \text{the mean of } M(C).$$

First let A be a MAX node. The mean m of $M(A)$ usually cannot be represented in a simple form in m_1 and m_2, the means of $M(B)$ and $M(C)$, respectively. Still, m is never less than the maximum of m_1 and m_2:

$$m \geq \max\{m_1, m_2\}.$$

Since m is usually larger than this maximum, m is often underestimated by the maximizing procedure alone. For an estimate of m relative to the estimates V_1 and V_2 of m_1 and m_2, the M & N procedure adds a bonus function to the maximum of V_1 and V_2 to improve the maximizing process:

$$V = \max\{V_1, V_2\} + \text{bonus function}.$$

In the case when A is a MIN node, the approximation becomes:

$$V = \min\{V_1, V_2\} - \text{bonus function}.$$

Furthermore, the bonus function depends on the distribution of the values at those nodes being considered. The bonus function increases if more nodes have the same maximum (or minimum) value. For example, one can consider the bonus function as being a function of the difference between V_1 and V_2, DELTA $= |V_1 - V_2|$. The most direct way to determine the proper function is to use experimentation. For the game of kalah, Slagle and Dixon (1970) used a bonus function of the following form:

$$1 + c \text{ DELTA},$$

where c is a negative value. In this bonus function, 1 is the maximum bonus value, and 0 is the minimum bonus value.

In general, the M & N procedure backs up all the M highest values of sons of a MAX node and all the N lowest values of sons of a MIN node. For the above function DELTA, $M = N = 2$. The conventional minimax procedure can also be treated as a special M & N procedure with $M = N = 1$. The M & N procedure has been demonstrated (Slagle and Dixon 1970) to be better than the minimax procedure for the game of kalah in the sense that the player using this new method won more games than an opponent using the conventional minimax procedure.

3.3.3 The *-MIN Procedure

Reibman and Ballard (1983a, b) proposed a new back-up procedure, called the *-MIN procedure. The main idea in this new procedure is to model the opponent's (MIN's) behavior explicitly. In the conventional minimaxing process, it is implicitly assumed that the opponent (MIN) is to minimize the same heuristic values as MAX is to maximize. These values are treated as if they were actual payoffs or actual minimax values.

In the *-MIN procedure, MIN is assumed to be fallible and to make decisions randomly according to a parameter P_s, called the *predicted strength*. This predicted strength is the probability that, given a choice of b moves, the opponent (MIN) will choose the nth best move over the $(n+1)^{th}$ best move. Therefore, the MIN nodes are treated as chance nodes, and the estimate at a MIN node is given by the following weighted average:

$$P_s M_1 + (1 - P_s) P_s M_2 + \ldots + (1 - P_s)^{b-1} P_s M_b,$$

where M_i is the expected value of the ith best move. At MAX nodes, the maximizing process is still used. The name *-MIN means that the MIN nodes are treated as chance nodes.

Instead of estimating the actual minimax value of a node, the *-MIN procedure estimates the final result of the game from each node. The experiments (Reibman and Ballard 1983b) showed that this approach is better for decision making than the minimax procedure when the predicted strength P_s is near MIN's actual behavior.

The formulation of this model is not precise. The order of good moves at MIN nodes is difficult to define. If the order depends on the actual minimax values of the children of the MIN node, then those minimax values should be exactly given, but this is generally assumed to be impossible in a heuristic search. Furthermore, the averaging process above holds only for a specifically randomized MIN, which is unlikely to occur. In fact, MIN's decision can also reasonably depend on MIN's own heuristic search, which may likely induce other decision making.

3.3.4 Average Propagation

The conventional minimaxing process is the right way to combine values at a node if those values are actual minimax values. It is also the best way to combine values if the player's opinions of the values of previously analyzed positions do not change on later moves, that is, if those values can be treated as actual values. However, real game-playing programs reanalyze positions after each move is made and usually come up with slightly different opinions on the later analyses.

The product-propagation procedure introduced in Sect. 3.3.1 is a correct way of estimating actual minimax values (for independent values) and is, therefore, the best way to combine values if they are estimates of probabilities of forced wins, and if no one is going to make any mistakes after the next move. For there to be no mistakes, after the next move the evaluation of nodes must be drastically changed to be perfect and actual minimax values must become the criteria of moves for both players.

An actual game-playing situation is generally somewhere between the two hypothetical extremes (static or drastically changed evaluations) described above. When an actual game-playing program eventually moves to a node, then the values computed at each move are progressively more accurate estimates of the node. Although the errors in these estimates decrease after each move, they usually do not drop to zero. Therefore, it should be better to use an approach which is between the two extremes of minimax propagation and product propagation.

Nau, Purdom, and Tzeng (1986) have thus used average propagation, which takes the average result of both minimax propagation and product propagation:

$$V_{\text{MAX}} = \frac{1}{2}\left(\max_i V_i + 1 - \prod_i \left(1 - V_i\right)\right)$$

for a MAX node, and

$$V_{\text{MIN}} = \frac{1}{2}\left(\min_i V_i + \prod_i V_i\right)$$

for a MIN node, where V_i ranges through the values of the children of the node. Although this average propagation was able to do better than both minimax and product propagations under many conditions (Nau, Purdom, and Tzeng 1986), which method of propagation works best in a given case still depends on both the static evaluation function and the game structure.

4 Probability Spaces and Martingales

In this book, a game space on which a heuristic-search model is constructed is always a probability space. All notions in the model, such as heuristic information, strength of nodes, and so on, are formulated in the terminology of probability theory. We use the concept of martingales to represent relationships among various node estimates that are based on heuristic information.

The concepts in probability theory needed in this book are quickly introduced in this chapter. Most results presented here do not have detailed discussions and proofs; these can be found in a standard book on probability theory (e.g., Chung 1974).

4.1 Borel Fields and Partitions

Let Ω be an abstract non-empty set of elements, called points and denoted by ω. Basic notation and the usual operations between subsets of Ω are as follows:

Element:	$\omega \in \Omega$
Subset:	$A \subseteq B$
Proper Subset:	$A \subset B$
Union:	$A \cup B$
Intersection:	$A \cap B$ or AB
Complement:	$A^C = \Omega \setminus A$
Difference:	$A \setminus B = A \cap B^C$

Usually, the set Ω is called a space. In this book, we consider spaces of games, called game spaces. A game space consists of a set of games which have the same game graph (or game tree). Each game in the game space can be represented by the list of values at the terminals of its game graph.

Consider the P_2-game space having the game tree in Fig. 2.4 as an example. Each game is constructed by assigning a 1 or a 0 independently to each entry of the 4×4 playing board. The probabilities of having a 1 and a 0 in an entry are p and $1 - p$, respectively. A game in this space is represented by its initial board configuration, a list of 16 numbers. Since each entry has two possible values, a total of $2^{16} = 65536$ games potentially exist in the space.

In a game space we consider special kinds of non-empty collections of subsets; such a collection is called a Borel field.

Definition 4.1. A non-empty collection F of subsets of a space Ω is called a Borel field on Ω if the following two closure properties hold

(1) If $A \in F$, then $A^C \in F$.

(2) If $A_j \in F$ for each integer $j\,(1 \leq j < \infty)$, then $\cup_j A_j \in F$

When there is no ambiguity, we use the expression "Borel field" to refer to a Borel field on the space under consideration. Condition (1) indicates that the complement of each element in a Borel field is also an element of the Borel field. Condition (2) also includes finite unions; that is, if A and B are elements of a Borel field F, then their union $A \cup B$ belongs to F, too. By using De Morgan's law,

$$(A \cup B)^C = A^C \cap B^C,$$

it can be shown that a Borel field F is also closed under finite intersection:

$$\text{If } A, B \in F, \text{ then } A \cap B \in F.$$

Furthermore, since $A \cap A^C = \emptyset$ and $A \cup A^C = \Omega$, each Borel field F consists of at least the empty set $\emptyset$ and the whole space Ω:

$$\emptyset \in F, \quad \Omega \in F.$$

If F_1 and F_2 are two Borel fields, and if each element of F_1 is also an element of F_2, $F_1 \subseteq F_2$, then F_1 is said to be a *subfield* of F_2. The collection $F_\emptyset = \{\emptyset, \Omega\}$, consisting of only the empty set $\emptyset$ and the whole space Ω, is the smallest Borel field in the sense that $F_\emptyset$ is a subfield of any Borel field on Ω. $F_\emptyset$ is called the *trivial* Borel field. On the other hand, the collection F_t of all subsets of Ω is also a Borel field. This is the largest Borel field and it is called the *total* Borel field; that is, any Borel field is a subfield of the total Borel field.

An element of a Borel field F is also called an *event* in F; therefore, an event is a subset of points. In game spaces, an event is usually a subset of games that have some common specific properties. In this book we use "event" to represent a piece of heuristic information about a game (Chap. 6).

In the discussion of a heuristic search, we consider the special Borel field that is generated by a partition of the space. Let $P = \{A_i\}\,(1 \leq i < m)$ be a finite or countable collection of subsets; i.e., m is a positive integer or $m = \infty$. P is called a *partition* of the space if the following two conditions hold:

(1) $\displaystyle\bigcup_{1 \leq i < m} A_i = \Omega$.

(2) $A_i \cap A_j = \emptyset$ if $1 \leq i, j < m$, and $i \neq j$.

Note that in this book a partition is always finite or countable; we do not consider uncountable partitions.

Given a partition $P = \{A_i\}\,(1 \leq i < m)$ consider the collection F_P of all possible unions of elements of P, including the empty set $\emptyset$, the union of no elements. From the properties of a partition, it can be easily verified that F_P is a Borel field, known as the Borel field generated by the partition P. If P consists only of elements of a Borel field F, then the generated Borel field F_P is a subfield of F.

Let $Q = \{B_j\}$ $(1 \leq j < n)$ also be a partition. Q is said to be *finer* than the partition $P = \{A_i\}$ $(1 \leq i < m)$, denoted by

$$P \leq Q,$$

if each $A_i (1 \leq i < m)$ is an element of F_Q, that is, if each A_i is a union of elements of Q. Q is also called a *refinement* of P. Therefore, the partition Q is finer than P if and only if F_P is a subfield of F_Q:

$$P \leq Q \quad \text{if and only if} \quad F_P \subseteq F_Q.$$

Note that in our definition, P itself can be a partition finer than P. If the partition Q is finer than P but not equal to P, then Q is said to be *properly* finer than P, or to be a proper refinement of P, denoted by

$$P < Q.$$

Q is properly finer than P if and only if F_P is a proper subfield of F_Q:

$$P < Q \quad \text{if and only if} \quad F_P \subset F_Q.$$

Consider again the P_2-game space with the game tree in Fig. 2.4 and study the heuristic search with the one-counter that counts the number of 1's in each node of the game tree.

At the root, the one-counter searches the whole initial playing board for the number of 1's, denoted by l. The value l ranges from 0 to 16, and there are 17 possible different values. For each l consider the subset of all games with exactly l 1's in the initial playing board. These 17 subsets form a partition of the whole game space.

In each partition element characterized by l, all games have the same heuristic information at the root relative to the one-counter; that is, there are l 1's in the root. Therefore, this heuristic search at the root corresponds to a partition of the whole space, and, for each game, the one-counter searches for the partition element to which the searched game belongs.

Consider the search at the two successors of the root, a search of the right and left halves of the playing board. Let the result be represented by a pair (l_1, l_2), l_1 being the number of 1's in the left half and l_2 being the number of 1's in the right half. Since there are 8 entries in each half, l_1 and l_2 both range from 0 to 8, and there are hence 81 different potential pairs in all.

For each pair (l_1, l_2), $0 \leq l_1$, $l_2 \leq 8$, let the subset of all games in which the numbers of 1's in the left and right halves of the initial playing board are l_1 and l_2, respectively, also be represented by the pair (l_1, l_2). Then any two different pairs correspond to two disjoint subsets, and there are 81 pairwise disjoint subsets in all. These 81 subsets form a partition of the whole game space. For each game, the one-counter, counting the number of 1's in each half of the playing board, determines the subset to which the searched game belongs.

Compare the heuristic searches at the root and at the two successors. For any game, let l and (l_1, l_2) be the corresponding results. Then it is necessary in this situation that

$$l = l_1 + l_2,$$

which indicates that the heuristic information at the root is always derivable from the heuristic information at the two successors. In other words, the one-counter has better visibility at the two successors than at the root.

Consider the two corresponding partitions. The above relation means that the subset represented by (l_1, l_2) is contained in the subset characterized by l. Moreover, for each l, the union of all subsets represented by (l_1, l_2) with $l = l_1 + l_2$ is the subset characterized by l. That is, the partition corresponding to the search at the two successors of the root is a proper refinement of the partition relative to the search at the root.

It is easy to extend the above discussion to the search at the next deeper level, at which the heuristic search corresponds to an even finer partition.

The formulation of heuristic searches and heuristic information in this book is based on the above discussion. For a game space, a heuristic search at a certain set of nodes in the game graph corresponds to a partition of the game space, and the searching process is designed to look for the partition element to which the searched game belongs. Heuristic information about a game can thus be characterized by this corresponding partition element. The fineness of partitions, then, can be used in formulating the visibility of the heuristic search; that is, a finer partition corresponds to a better visibility.

The Borel fields generated by partitions in a heuristic search are essential for developing a theory for estimating nodes based on heuristic information. Another essential concept, a probability measure on a Borel field, is introduced in the next section.

4.2 Probability Spaces

Let Ω be a space, and let F be a Borel field on Ω. A probability measure is defined as follows.

Definition 4.2. A *probability measure* P on F is a numerically valued set function with domain F, satisfying the following axioms:

(1) $P(A) \geq 0$ for any $A \in F$.
(2) If $\{A_i\}$ is a countable collection of pairwise disjoint sets in F, then $P(\cup_i A_i) = \Sigma_i P(A_i)$.
(3) $P(\Omega) = 1$.

Let P be a probability measure on F; then the triple (Ω, P, F) is called a *probability space*. For each element A in F, the value $P(A)$ is called the probability measure (or measure) of A. Therefore, any element of F having a measure in this sense is also called a *measurable set* (relative to F). Note that the term "events" also denotes elements of F when we emphasize that the corresponding elements refer to heuristic information about a game. If P and F are given, and if there is no danger of ambiguity, we also call the space Ω itself a probability space.

Axiom (2) is called *countable additivity*, which implies finite additivity; that is,

$$P(A \cup B) = P(A) + P(B)$$

if A and B are two disjoint measurable sets.

If Ω is a finite or countable set and if each singleton containing only one point is measurable, then the space is said to be *discrete*. Let $\Omega = \{X_i\}$ $(1 \leq i < n)$ be discrete, and let $P(X_i)$ be the measure of the singleton containing X_i. Then any subset is measurable, and its measure is the sum of the measures of all points in the subset. For the whole space, the sum is always 1:

$$\sum_{1 \leq i < n} P(X_i) = 1.$$

Consider again the P_2-game space with the game tree in Fig. 2.4. Each entry on the initial playing board independently assumes the values 1 or 0 with probabilities p and $1 - p$, respectively $(0 \leq p \leq 1)$. For an arbitrary game, let the number of 1's on the playing board be l. Then the number of 0's on the playing board is $16 - l$ and the probability of this board configuration is

$$p^l (1 - p)^{16 - l}.$$

Using this value to denote the measure of the game, the whole P_2-game space becomes a discrete probability space. Any set of games is measurable, and its measure is the sum of the probabilities of all games in the set. For example, let A_l be the event (a measurable set) consisting of all games with exactly l 1's on the playing board. Since the number of games in A_l is the binomial coefficient

$$\binom{16}{l},$$

the measure of A_l is

$$P(A_l) = \binom{16}{l} p^l (1 - p)^{16 - l}.$$

The search with the one-counter at the two successors of the root returns a pair (l_1, l_2), in which l_1 is the number of 1's in the left half and l_2 is the number of 1's in the right half of the playing board. Let $A_{(l_1, l_2)}$ be the corresponding event. Since there are

$$\binom{8}{l_1} \cdot \binom{8}{l_2}$$

games in this event, the measure of this event is

$$P(A_{(l_1, l_2)}) = \binom{8}{l_1} \binom{8}{l_2} p^{l_1 + l_2} (1 - p)^{16 - l_1 - l_2}.$$

A game space with a probability measure is called, in this book, a probabilistic game space.

4.3   Random Variables

Let (Ω, F, P) be a probability space. In probability theory a random variable is generally extended-valued; that is, both $+\infty$ and $-\infty$ can be the values of a random variable. And the domain of a random variable is generally only a subset of Ω. However, in this section we consider real, finite-valued random variables on the whole space Ω.

Let $\mathbb{R} = (-\infty, \infty)$ be the finite real line and $\mathbb{B}$ the Borel field on $\mathbb{R}$ generated by all open intervals. An element of $\mathbb{B}$ is said to be a *linear Borel set*. Then a random variable is defined as follows.

Definition 4.3.   A real and finite-valued random variable X is a function whose domain is Ω and whose range is contained in $\mathbb{R}$ such that, for each $B \in \mathbb{B}$,

$$\{\omega | X(\omega) \in B\} \in F.$$

The above notation for an element in F is also abbreviated to $\{X \in B\}$. Consider the inverse mapping X^{-1} from $\mathbb{R}$ to Ω, defined as follows:

$$X^{-1}(A) = \{\omega | X(\omega) \in A\} = \{X \in A\} \text{ for all } A \subset \mathbb{R}.$$

Then the condition of X being a random variable is equivalent to the fact that X^{-1} maps members of $\mathbb{B}$ onto members of F:

$$X^{-1}(B) \in F \text{ for all } B \in \mathbb{B}.$$

That is, the inverse image of a measurable set in $\mathbb{R}$ (relative to $\mathbb{B}$) is measurable in Ω (relative to F). Such a function is also called a *measurable function*. Therefore, a random variable is the same as a measurable function from Ω to $\mathbb{R}$.

On a game space, the most common functions are (1) the minimax values and (2) the game values, relative to players' strategies, introduced in this book. For the purpose of our theoretical study, we shall always assume that all of the values under consideration in a probabilistic game space are random variables. If the game space is discrete, then any real, finite-valued functions are random variables, including minimax values and game values, because any subsets of games are measurable.

In computer science, averaging is a very commonly used analytic technique for estimating functions. For example, the averages of some relevant values are important for studying both the complexity and the performance of an algorithm. In the terminology of probability theory, averaging is finding the mean of a random variable. The mean is also called the *expectation* of the random variable.

In a probability space (Ω, P, F), a concept of integration with respect to the probability measure can be introduced, similar to the usual integration of real-valued functions on the real line. Let f be a random variable, and let A be a measurable set. Then the integral of f over A is denoted by

$$\int_A f \, dP.$$

Without giving the definition of integration in detail, we list here some properties needed in our discussion.

A random variable f is said to be integrable over the whole space Ω if the integral

$$\int_\Omega f \, dp$$

is finite; this finite integral is called the expectation of f. The expectation of f, denoted by $E(f)$, is also called the *mean* or the average of f. If f is integrable, then the integral of f over any measurable set A is also finite.

In a discrete space $\Omega = \{\omega_i\} \, (0 \le i < n)$ the integral of a random variable f is the following summation:

$$E(f) = \sum_{0 \le i < n} P(\omega_i) f(\omega_i).$$

Therefore, any random variable in a finite discrete space is integrable. Moreover, if each point in the finite space has the same probability, the mean of a random variable becomes the usual average:

$$E(f) = \frac{1}{n} \sum_{0 \le i < n} f(\omega_i).$$

The above two equations are the most commonly used techniques for finding averages in computer science.

Next we introduce the fundamental concept of independence.

Definition 4.4.   The random variables $\{X_i\} \, (1 \le i \le n < \infty)$ are said to be independent if for any linear Borel sets $\{B_i\}$, $(1 \le i \le n)$ (i.e., elements of $\mathbb{B}$),

$$P\left(\bigcap_{i=1}^{n} (X_i \in B_i) \right) = \prod_{i=1}^{n} P(X_i \in B_i).$$

Definition 4.5.   The events $\{E_j\}$ $(1 \le j \le n)$ are said to be independent if, for any subset $\{j_1, \ldots, j_l\}$ of $\{1, \ldots, n\}$,

$$P\left(\bigcap_{k=1}^{l} E_{j_k} \right) = \prod_{k=1}^{l} P(E_{j_k}).$$

The most important property of independent random variables is the following theorem. (See Chung (1974) for a proof.)

Theorem 4.1.   If X and Y are independent and both have finite means, then

$$E(XY) = E(X) E(Y).$$

This result can be extended to a finite number of random variables:

Corollary.   If $\{X_i\}$ $(1 \le i \le n)$ are independent random·variables with finite means, then

$$E\left(\prod_{i=1}^{n} X_i \right) = \prod_{i=1}^{n} E(X_i).$$

4.4   Product Spaces

Let (Ω_j, F_j, P_j), $1 \le j \le n$, be n probability spaces, n being a positive integer. The product space

$$\Omega = \prod_{j=1}^{n} \Omega_j$$

is the collection of all ordered n-tuples

$$\omega = (\omega_1, \ldots, \omega_n),$$

where for each j $(1 \le j \le n)$ $\omega_j \in \Omega_j$.

Any subset A of Ω is said to be a *product set* if it is a product of measurable sets:

$$A = \prod_{i=1}^{n} A_j, \quad A_j \in F_j \ (1 \le j \le n).$$

Let F be the Borel field generated by all the product sets; F is the smallest Borel field containing any product sets. Then F is called the *product Borel field* of $\{F_j\}$ $(1 \le j \le n)$ and is denoted by

$$F = \prod_{i=1}^{n} F_j.$$

A natural probability measure P on the product Borel field F is such that for each product set,

$$P(A) = \prod_{j=1}^{n} P_j(A_j).$$

This probability measure P is called the *product measure* of $\{P_j\}$ $(1 \le j < n)$ and is denoted by

$$P = \prod_{j=1}^{n} P_j.$$

This new probability space (Ω, F, P) is called the *product probability space* of (Ω_j, F_j, P_j), $1 \le j \le n$.

Consider again the P_2-game space with the game tree in Fig. 2.4 as an example. Since each entry on the playing board (independently) assumes the value 1 or 0 with probabilities p or $1 - p$, respectively, the value space $\{0, 1\}$ in each entry forms a probability space in which the probability of 1 is p and the probability of 0 is $1 - p$. Since there are 16 entries on the original playing board, the game space of all possible games is a product space

$$\Omega = \{0, 1\}^{16}.$$

The probability measure used so far is just the product probability. That is, the probability of a board configuration with exactly l 1's and $(16-l)$ 0's in previously specified entries is

$$p^l(1-p)^{16-l}.$$

Similarly, general P_b- and G_d-game spaces are also product probability spaces.

Let (Ω, F, P) be the product probability space of (Ω_j, F_j, P_j), $1 \leq j \leq n$. For each $j (1 \leq j \leq n)$ let X_j be a random variable on Ω_j. X_j induces a function X_j' on Ω, defined by

$$X_j'(\omega) = X_j(\omega_j), \quad \omega = (\omega_1, \ldots, \omega_j, \ldots, \omega_n).$$

Then X_j' is a random variable on Ω and the $\{X_j'\}$ $(1 \leq j \leq n)$ are independent. To simplify our use of symbols, we also use X_j to represent the induced X_j'. For example, the minimax value of a node in the game tree of a P_2-game space, being a function of the leaves under the node, is also treated as a function of the whole collection of leaves under the root. However, only the values at the leaves under the node decide the minimax value, and the values at other leaves do not affect this value.

4.5   Conditional Probabilities and Martingales

Let (Ω, F, P) be a probability space, and let A be an element of F with $P(A) > 0$. The conditional probability relative to A is defined as follows.

Definition 4.6.   For each measurable set $B \in F$ the value

$$P(B|A) = \frac{P(A \cap B)}{P(A)}$$

is called the *conditional probability* of B, given A.

Consider again the P_2-game space with the game tree in Fig. 2.4. Suppose it is given, for a random game in the space, that there are 10 ones on the playing board. What is the probability that there are 8 ones on the right half of the playing board? Let A be the event of games with 10 ones on the playing board. Let B be the event of games with 8 ones on the right half of the playing board. Then our problem is to find $P(B|A)$, the conditional probability of B, given A.

Note that the intersection $A \cap B$ (or AB) is the event of games with 8 ones on the right half and 2 (i.e., $10-8$) ones on the left half of the playing board. Therefore,

$$P(A \cap B) = \binom{8}{8} p^8 \binom{8}{2} p^2 (1-p)^{8-2} = \binom{8}{2} p^{10}(1-p)^6.$$

Since

$$P(A) = \binom{16}{10} p^{10}(1-p)^6,$$

the conditional probability of B, given A, is

$$P(B|A) = \frac{\binom{8}{2}}{\binom{16}{10}} = \frac{1}{286} = 0.0035.$$

We now return to a general probability space (Ω, F, P). One application of conditional probability is the so-called *Baye's rule*, which is the starting point for an entire statistical philosophy known as Bayesian statistics. Let $\{A_i\}$ $(1 \leq i \leq n)$ be a partition of the space Ω such that $P(A_i) > 0$ for each i, and let B be an event for which $P(B) > 0$. Then we have the following two relations:

$$P(A_j|B) = \frac{P(A_j \cap B)}{P(B)} = \frac{P(A_j)P(B|A_j)}{P(B)}$$

and

$$P(B) = \sum_{i=1}^{n} P(A_i \cap B) = \sum_{i=1}^{n} P(A_i)P(B|A_i).$$

Combining the above relations we get the famous Baye's rule:

$$P(A_j|B) = \frac{P(A_j)P(B|A_j)}{\sum_{i=1}^{n} P(A_i)P(B|A_i)} \quad (1 \leq j \leq n).$$

Given a measurable set A with $P(A) > 0$, consider a set function P_A on the Borel field F, defined as:

$$P_A(B) = P(B|A), \quad B \in F.$$

It can be shown that P_A is also a probability measure on F. If X is an integrable random variable, then the expectation of X relative to P_A, denoted by $E_A(X)$, is

$$E_A(X) = \int_{\Omega} X dP_A = \frac{1}{P(A)} \int_A X dP.$$

$E_A(X)$ is called the *conditional expectation* of X relative to A.

Once again let $\{A_i\}$ $(1 \leq i \leq n)$ be a partition of the space Ω such that $P(A_i) > 0$ for each i, and let F_1 be the Borel subfield generated by this partition. For an integrable random variable X, we define a new function, denoted by $E_{F_1}(X)$, as follows:

$$E_{F_1}(X) = \sum_{i=1}^{n} E_{A_i}(X) \mathbf{1}_{A_i},$$

where 1_{A_i} is the indicator of A_i:

$$1_{A_i}(\omega) = \begin{cases} 1 & \text{if } \omega \in A_i; \\ 0 & \text{otherwise.} \end{cases}$$

$E_{F_1}(X)$ is a discrete random variable because for each $i(1 \le i \le n)$:

$$E_{F_1}(X)(\omega) = E_{A_i}(X), \quad \omega \in A_i.$$

F_1 is the Borel subfield generated by A_i; therefore, $E_{F_1}(X)$ is also measurable relative to F_1. The expectation of this new function is

$$\int_\Omega E_{F_1}(X)\,dP = \sum_{i=1}^n \int_{A_i} E_{F_1}(X)\,dP = \sum_{i=1}^n P(A_i) E_{A_i}(X) = E(X);$$

that is, X and $E_{F_1}(X)$ have the same expectation. Generally, for any measurable set A in F_1, namely, a union of a subcollection of A_i's, a similar consideration yields that

(*)
$$\int_A X\,dP = \int_A E_{F_1}(X)\,dP.$$

In fact, the function $E_{F_1}(X)$ is uniquely characterized by the properties of being measurable relative to F_1 and satisfying the relation (*) for any measurable set $A \in F_1$. $E_{F_1}(X)$, called the *conditional expectation* of X relative to F_1, is generally defined as follows.

Definition 4.7. Given an integrable random variable X and a Borel subfield F_1, the conditional expectation of X relative to F_1, denoted by $E_{F_1}(X)$, is any random variable satisfying the following two properties:

(1) It is measurable with respect to F_1.
(2) It has the same integral as X over any set in F_1; that is, the equation (*) holds for any $A \in F_1$.

The conditional expectation of an integrable random variable always exists, and it is uniquely determined in the sense that any two conditional expectations can differ only in a set of zero probability. Two random variables which differ only in a set of zero probability are said to be equal almost everywhere. In this book we adopt this generalized equality; that is, two random variables are said to be equal if they are equal everywhere or almost everywhere.

If F_1 is the trivial Borel field $\{\varnothing, \Omega\}$, then the conditional expectation is the constant function with the mean (i.e., the expectation) of X, $E(X)$, as the value

$$E_{\{\varnothing, \Omega\}}(X) = E(X)1_\Omega.$$

Relative to different Borel subfields, the conditional expectations have the following important properties.

Theorem 4.2. Let F_1 and F_2 be two Borel subfields of F such that $F_1 \subseteq F_2$, and let X be an integrable random variable. Then

$$E_{F_1}(X) = E_{F_2}(X)$$

if and only if $E_{F_2}(X)$ is also measurable relative to F_1. In general,

$$E_{F_1}(E_{F_2}(X)) = E_{F_1}(X) = E_{F_2}(E_{F_1}(X)).$$

Now we introduce the concept of a martingale.

Definition 4.8. A sequence (finite or infinite) of random variables and Borel subfields $\{X_i, F_i\}$ is called a *martingale* if, for each i,

(1) $F_i \subseteq F_{i+1}$, and X_i is measurable relative to F_i;
(2) X_i is integrable; and
(3) $X_i = E_{F_i}(X_{i+1})$.

Let $\{X_i, F_i\}$ be a martingale. From Theorem 4.2 we can derive the following relation:

$$E_{F_i}(X_j) = X_i \, (i < j).$$

Therefore, for each $A \in F_i \, (i \le j)$,

$$\int_A X_i \, dP = \int_A X_j \, dP.$$

Suppose that we have an increasing sequence of Borel subfields $\{F_i\}$ and an integrable random variable X. Let X_i be the conditional expectation of X relative to F_i:

$$X_i = E_{F_i}(X).$$

Then, from Theorem 4.2, the sequence $\{X_i, F_i\}$ forms a martingale.

Historically, the idea of martingale has been used to formulate a system of gambling. However, we use a martingale to represent a system of estimating a random variable. Let X be a random variable, such as the minimax value or a game value of a node in a game tree. Suppose that F_i is a Borel subfield generated by the partition of a game space corresponding to a heuristic search. Then we treat $X_i = E_{F_i}(X)$ as an estimator of X. If $\{F_i\}$ is an increasing sequence of Borel subfields, then $\{X_i, F_i\}$ becomes a martingale. This means that X_{i+1} is a more precise estimator of X than is X_i. Such a model for estimating random variables is discussed fully in Chap. 7.

5   Probabilistic Game Models and Game Values

In this chapter we introduce a formal definition of a probabilistic game model for further theoretical study. The P_2- and G_1-games introduced in Chap. 2 are two probabilistic game models. A probabilistic game model is a space of zero-sum two-person perfect information games such that all of the games in the space have the same fixed game graph (or tree) and the values on leaves are assigned according to a fixed distribution. Because of this probabilistic assignment of the values at leaves, a game in a probabilistic game model can be treated as a random event in a probabilistic space.

A notion of strategies describing the decision behaviors of the game players (MAX and MIN) is also introduced here. Based on pairs consisting of a strategy of MAX and a strategy of MIN, a (relative) game value can be assigned to each node in the game graph to represent the expected result of the subgame starting from the node. In this case, the minimax values are special game values relative to the minimax optimal strategies. Game values, being random variables, are used to represent varied kinds of node strength, and estimating the strength of a node is thus equivalent to estimating the corresponding game value or random variable.

Finally, two special modes, P_b- and G_d-game models, are introduced, and the properties of minimax values in both models are also discussed. Based on these properties, we discuss later (Chaps. 9 and 10) the estimation of these minimax values.

5.1   Probabilistic Game Models

Let T be a game graph, as introduced in Chap. 2. A game graph T is characterized by the following properties: T is a finite connected directed graph with an explicitly given node called the *root* being the only node without predecessors. The nodes without successors are called leaves or *terminal* nodes. Any path starting from the root terminates at a terminal node in a finite number of edges; that is, there is no loop in T. There are two types of nodes in T, MAX nodes and MIN nodes; they alternate along any path from the root to a terminal node.

As in Chap. 2, the height of a node in a game graph T is defined as the length of the longest path from that node to a terminal node. Also, if T is a tree, nodes belonging to the set of all nodes reachable from the root in the same number of edges are said to have the same level.

Given the game graph T, we can construct a finite zero-sum two-person perfect information game by assigning each terminal node a (real) value. Call the two players MAX and MIN. MAX moves from MAX nodes, and MIN moves from MIN nodes. The goal of the game is that MAX wishes to maximize the values at terminals and MIN wishes to minimize them. Such a game is called a T-game.

Suppose in the game graph T there are $n(>0)$ terminal nodes, previously arranged in a specific order. Then a T-game can be represented by a vector of n components, $X = (X_1, \ldots, X_n)$, where X_i is the value at the i-th terminal node. Hereafter, we use such a vector X to represent a T-game.

For the game graph T, a model of games is a set of T-games. A probabilistic game model is a game model with a probability distribution; a T-game in this model is set up randomly according to this probability distribution. Formally, a probabilistic game model is defined as follows.

Definition 5.1. A probabilistic game model for the game graph T is a probability space (Ω, F, P), where Ω is a sample space of T-games, F is a Borel field of subsets of Ω, and P is a probability measure on Ω relative to F.

The triple (Ω, F, P) is used to represent the corresponding probabilistic game model if there is no ambiguity about the game graph T. For example, the P_2-game model for the game tree in Fig. 2.4 is a probabilistic model, in which the space $\{0, 1\}^{16}$ is the sample space with the probability measure induced by the probabilities of 0 and 1 at each terminal node. In this model there are a total of $2^{16} = 65536$ possible games. The G_1-game model for the game graph in Fig. 2.2 is another probabilistic game model, in which the sample space is $\{0, 1\}^5$. There are only $2^5 = 32$ potential games in this model.

5.2   Strategies and Game Values

Let (Ω, F, P) be a probabilistic game model for a game graph T, and let X be a T-game assigned to the game graph T. Then the T-game X is a random vector. Now we consider possible ways the two players MAX and MIN can play a T-game X.

First we introduce the concept of strategies (Dresher 1981, Tzeng 1984). In an actual playing of the game, each player may formulate in advance a plan for playing the game from beginning to end. Such a plan must be complete and cover all possible contingencies that may arise during playing of the game. Among other things, this plan might incorporate any information which may become available to the player in accordance with the rules of the game. For example, MAX could use the following simple plan: to search one step ahead from each of his positions with a certain evaluation function and then to choose the node with the largest searched value for the next move. At each step a player could also consider the opponent's previous values. A complete plan of a player like this is called a strategy.

5.2.1   Non-randomized Strategies

Formally, we first define a non-randomized strategy, a strategy for choosing a unique move from each possible position.

Definition 5.2. A *non-randomized strategy* of a player (MAX or MIN) is a function f such that, for any T-game X in the model and any node N of the player in T, $f(X, N)$ is a successor of the node N.

Note that, in this definition, the node N is a MAX node if the player is MAX; otherwise, it is a MIN node. A player is said to play the game with a strategy f if, for any given T-game X, he chooses the node $f(X, N)$ from any node N from which he is to move. In other words, the strategy f completely determines a non-randomized decision-making process for playing games in the model. If we consider the game tree, instead of the game graph, then a strategy can also take previous moves into account. In a game tree, the moves from the root to a node are uniquely determined. Therefore, a strategy can also be treated as a function of a game and previous moves.

In order to discuss possible results of a game, we consider a complete round, that is, a sequence of moves from beginning to end. A complete round is represented by a path in the game graph T from the root to a terminal node: $(A_1, \ldots, A_h)$, where A_1 is the root, A_{i+1} is a successor of A_i ($0 < i < h$), and A_h is a terminal node. If the subgame from a node N in T is played, then the first node of a path $(A_1, \ldots A_h)$ representing a complete round of this subgame is the node N: $A_1 = N$. Note that h is the number of nodes in this entire path and that h may be different for different rounds of the same game.

A path representing a complete round of a subgame or the whole game is called a *playing path* of the subgame or the whole game. The value at the terminal node of a playing path is the result of the corresponding round. Therefore, a playing path provides both the history and the result of a round.

Let S_{MAX} and S_{MIN} denote the set of all non-randomized strategies of MAX and MIN, respectively. Since a player's strategy completely determines his decision making, from any node N in T and for a T-game X, each pair (f, g) of strategies, $f \in S_{\text{MAX}}$ and $g \in S_{\text{MIN}}$, determines a unique playing path of the subgame from N: $(A_1, \ldots, A_h)$. In this path, $A_1 = N$; for each i ($0 < i < h$), $f(X, A_i) = A_{i+1}$ if A_i is a MAX node, and $g(X, A_i) = A_{i+1}$ if A_i is a MIN node. If N is the root of T, then the playing path is for the whole game. Such a path is called a playing path from N relative to the strategies f and g. Now we define the game values as follows.

Definition 5.3. For a T-game X, let $(A_1, \ldots A_h)$ be the playing path from a node N relative to strategies $f \in S_{\text{MAX}}$ and $g \in S_{\text{MIN}}$. Then the value at the terminal node A_h is called the *game value* at the node N with respect to f and g and is denoted by $M_N^{(f, g)}(X)$.

Suppose that MAX and MIN play the game from a node N, using the strategies f and g, respectively. Given any T-game X, then the game value $M_N^{(f, g)}(X)$ is the result of this particular round. Therefore, for any node N in T and any pair of $f \in S_{\text{MAX}}$ and $g \in S_{\text{MIN}}$, we have a function

$$M_N^{(f,\,g)}: \Omega \rightarrow D,$$

where D is the domain of payoffs of the games, i.e., the set of all possible values at terminal nodes. For the purpose of an analytic study, we always assume that the game value $M_N^{(f,\,g)}(X)$ is a random variable for any $N, f,$ and g.

5.2.2 Randomized Strategies

In this section we generalize the concept of strategies to include randomized strategies, which are defined as follows.

Definition 5.4. A *randomized strategy* of a player (MAX or MIN) is a multi-valued function which, for each T-game X and each node N of the player in the game graph T with successors $N_1, \ldots, N_s$, returns a value p_i to N_i for each i $(1 \leq i \leq s)$ so that $0 \leq p_i \leq 1$ for $1 \leq i \leq s$ and $\Sigma_{i=1}^s p_i = 1$.

The set of values $p_1, \ldots, p_s$ in the definition represents a probability distribution on the successors of the node N. A player is said to play the game with a randomized strategy f if, for any given T-game, he makes a randomized decision from each position according to the probability distribution returned by f at that position. Each non-randomized strategy introduced previously can now be treated as a special randomized strategy, which gives the chosen successor from a node the value 1 and gives other successors the value 0. Let R_{MAX} and R_{MIN} be the set of all randomized strategies of MAX and MIN, respectively. Then S_{MAX} is a subset of R_{MAX}, and S_{MIN} is a subset of R_{MIN}.

As in non-randomized cases, game values relative to randomized strategies can also be defined. Let $f \in R_{\text{MAX}}$, $g \in R_{\text{MIN}}$, and let N be a node in T. For a given T-game X, a value $M_N^{(f,\,g)}(X)$ is defined recursively as follows.

Definition 5.5. If N is the i-th terminal node and $X = (X_1, \ldots, X_n)$, then $M_N^{(f,\,g)}(X)$ is the value at the terminal node:

$$M_N^{(f,\,g)}(X) = X_i.$$

If N has $N_1, \ldots, N_s$ as its successors, then

$$M_N^{(f,\,g)}(X) = \sum_{i=1}^s p_i M_{N_i}^{(f,\,g)}(X),$$

where $p_1, \ldots, p_s$ are the values returned from the node N by the strategy f or g according to whether N is a MAX or a MIN node.

Suppose that MAX and MIN play an arbitrarily given T-game X from a node N, using randomized strategies f and g, respectively. Then the actual result of the game is generally no longer fixed (as it would be in non-randomized cases) because at each step the player (MAX or MIN) makes a randomized, instead of a fixed, move. In fact, the result is a new random variable of which the expectation is the game value $M_N^{(f, g)}(x)$. That is, if, from the node N, MAX and MIN play the same game X many times, using the same strategies f and g, the expected average of the final payoffs is $M_N^{(f, g)}(x)$.

$M_N^{(f, g)}(X)$ is also called the *game value* at the node N with respect to f and g for the T-game X. If f and g are non-randomized, then this definition is consistent with the previous one. Similarly, all game values are random variables in our game model.

5.2.3 Minimax Values

Next, we show that the conventional minimax values are special game values with respect to special strategies. First we recall the definition of minimax values. Let $X = (X_1, \ldots, X_n)$ be a given T-game in our probabilistic model, and let N be an arbitrary node in the game graph T. Then the minimax value, $M_N(X)$, of the node N is defined recursively as follows:

(1) If N is the i-th terminal node, then $M_N(X) = X_i$.
(2) If N has successors $N_1, \ldots, N_s$, then

$$M_N(X) = \max_i M_{N_i}(X)$$

in the case when N is a MAX node, and

$$M_N(X) = \min_i M_{N_i}(X)$$

in the case when N is a MIN node.

Having defined the minimax value of a node in T, we can consider a strategy which chooses from a node a successor having the same minimax value as the node itself. For MAX, such a strategy is characterized by choosing from a MAX node a successor with the largest minimax value. For MIN, it is characterized by choosing from any MIN node a successor with the smallest minimax value. If more than one successor has this largest or smallest value, an equal probability can be assigned to each such successor. The strategy described here is called in this book a *minimax optimal strategy*. Let f^* be a minimax optimal strategy of MAX, and let g^* be a minimax optimal strategy of MIN. Then the game value with respect to f^* and g^* is the minimax value itself:

$$M_N^{(f^*, g^*)}(X) = M_N(X)$$

for any node N and any T-game X in the model. That is, the minimax values are game values with respect to minimax optimal strategies.

It is easy to show by induction that the minimax optimal strategies of both players are characterized by the following relations:

$$M_N(X) = M_N^{(f^*, g^*)}(X)$$
$$= \max_f \min_g M_N^{(f, g)}(X)$$
$$= \min_g \max_f M_N^{(f, g)}(X)$$

for any node N in T and any T-game X in the model. Because minimax values are important in describing games, the finding of minimax values and minimax optimal strategies is a main topic in game theory.

5.3 P_b-Game Models

Let $b \geq 2$ be an integer, and let p be a real number $(0 \leq p \leq 1)$. Consider a probabilistic game model in which the game graph is a complete, uniform tree with the branching factor b and in which the value 1 (a win for MAX) or 0 (a win for MIN) is independently assigned to each terminal node with a probability p or $1 - p$, respectively. Such a probabilistic game model is called a P_b-game model. The P_2-game models introduced in Chap. 2 are special cases of P_b-game models with $b = 2$.

The P_b-game models are standard models used for theoretical study (Pearl 1980, 1984). Here we discuss the relationships between and the asymptotic behaviors of the means of the minimax values in these models. For simplicity, we assume that MIN is always the last player, the one with the last turn (at height 1) to move. That is, the terminal nodes (of height 0) and the nodes of height $2n(n > 0)$ are MAX nodes; the height of any MIN node is an odd number.

Given a game in a P_b-game model, let M_n be the random variable representing the minimax value of an arbitrary node of height n. Since the game tree is complete and uniform, and since the values (1 and 0) at the terminal nodes are assigned independently, according to the same probability distribution (p and $1 - p$), the minimax values of all nodes at the same level are independent and have an identical probability function. That is, they are independent and identically distributed (i.i.d.) random variables. Let E_n be the expected value (i.e., mean) of M_n. Then

$$E_n = 1 \cdot P(M_n = 1) + 0 \cdot P(M_n = 0).$$

$P(M_n = 1)$ is the probability that $M_n = 1$, and $P(M_n = 0)$ is the probability that $M_n = 0$. Thus, E_n is the probability that $M_n = 1$, that is, the probability that MAX wins the game starting from a node of height n if both players use minimax optimal strategies.

Since M_0 is the value at a terminal node, its expected value is the probability of 1 at the terminal node:

$$E_0 = p.$$

Let $n > 0$ be an even integer. Then, M_n is the minimax value of a MAX node; $M_n = 0$ if and only if the minimax value of each successor of the node is zero. Therefore, we have the following probability that $M_n = 0$:

$$1 - E_n = (1 - E_{n-1})^b.$$

That is,

$$E_n = 1 - (1 - E_{n-1})^b$$

if n is even.

Now let $n > 0$ be an odd integer. Then, M_n is the minimax value of a MIN node;

$M_n = 1$ if and only if the minimax value of each successor of the node is one. Therefore, the probability that $M_n = 1$ is

$$E_n = (E_{n-1})^b$$

if n is odd.

To simplify the notation, let

$$P_n = E_{2n} \quad \text{and} \quad Q_n = E_{2n+1}$$

for all $n > 0$. Then we have the following recurrence relations for P_n and Q_n:

$$P_0 = p;$$

$$P_n = 1 - (1 - P_{n-1}^b)^b; \quad \text{and}$$

$$Q_n = P_n^b.$$

The following discussions are based on the theorems presented in the appendix. For each $b > 1$, there is a value W_b in the interval $(0, 1)$ such that, if the original probability $p = W_b$, then

$$P_n = W_b \quad \text{and} \quad Q_n = W_b^b = 1 - W_b$$

for all n. That is, if the initial probability of 1 at the leaves is W_b, then the expected value of the minimax value of any MAX node is also W_b, independent of the height of the node. The expected value of the minimax value of any MIN node is $W_b^b = 1 - W_b$. No matter how deep the game tree is, if both players use minimax optimal strategies, MAX always has the same probability W_b of winning the game.

If $0 < p < W_b$, then $\{P_n\}$ is a strictly decreasing sequence and converges to 0. In other words, the game becomes more favorable to MIN (the last player) if the depth of the game tree increases and, asymptotically, MIN can win the game with probability 1. On the other hand, if $W_b < p < 1$, then $\{P_n\}$ is a strictly increasing sequence and converges to 1. That is, the game becomes more favorable to MAX if the depth of the game tree increases and, asymptotically, MAX can win the game with the probability 1. Both cases hold true only under the theoretical assumption that both MAX and MIN are using minimax optimal strategies. Practically, however, it is not feasible to find minimax optimal strategies for large game trees, and the above results can hardly represent actual asymptotical behavior.

5.4 G_d-Game Models

Let $d > 0$ be an arbitrary positive integer. The games discussed in this section have a row of $hd + 1$ squares as the playing board, where $h > 0$ is also a positive integer. The playing board is set up by independently assigning to each square the value 1 or 0, with probabilities p and $1 - p$ ($0 \leq p \leq 1$), respectively. A move (for either player) consists of removing, from one or both ends of the playing board, a total of d squares. There are exactly $d + 1$ different alternatives in each move; that is, a player can remove i squares from one end and $(d-i)$ squares from the other end for $i=0,\ldots,d$. The game ends when only one square is left; MAX wins if this square has the value 1, and MIN wins otherwise. Such a game is called a G_d-*game*. The G_1-games introduced in Chap. 2 are special cases of G_d-games with $d=1$.

In a G_d-game graph, suppose that the node representing a game position is labeled by the list of 1's and 0's on the squares of this game position, as illustrated in Fig. 5.1 for a G_2-game $(d=2)$. A general G_d-game graph is so organized that we have the following properties. From any given internal node having $d+1$ successors, if we remove all of the d entries in one move from the right end of the corresponding list, then we get the first successor from the left. If we remove $i(1\le i\le d)$ entries from the left end and remove $d-i$ entries from the right end, then we get the $(i+1)$th successor from the left. Furthermore, if we remove d entries from each end of the list, then we get the only common successor of the $d+1$ sons of the given node, the middle grandson of this given node.

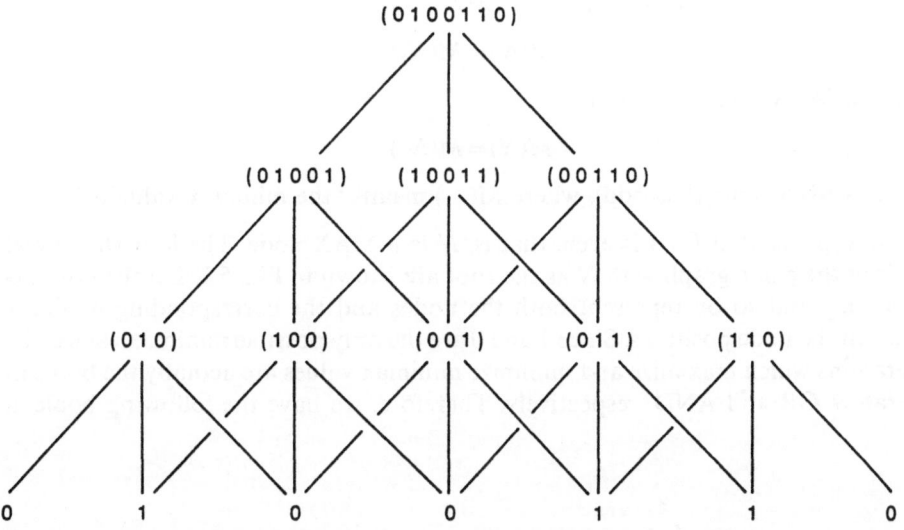

Fig. 5.1. The game graph of a G_2-game, where each node has three successors and each two adjacent siblings have two common successors

The levels of nodes in a G_d-game graph can be defined as follows. The root itself is the first level; its $d+1$ successors form the second level; and so on. The number of nodes at a level is increased by d at each next deeper level; therefore, at the ith level, there are $(i-1)d+1$ nodes in all. At each internal level, two consecutive nodes have d common successors. In general, $i(2\le i\le d+1)$ consecutive neighboring nodes have $d-i+2$ common successors. Specifically, $d+1$ consecutive neighboring nodes have exactly one common successor.

Let T be a G_d-game graph. Given a real value $p(0<p<1)$, a G_d-*game model* for T is now formally defined as the set of all G_d-games for which T is the corresponding game graph. Further, the values at terminal nodes are independently assigned with a probability p of 1 and a probability $1-p$ of 0.

Given a G_d-game model, without loss of generality, we assume that MIN is the last player. Therefore, the nodes with odd heights are MIN nodes, and the nodes with even heights are MAX nodes. Minimax values of nodes at different levels have a simple relationship. For the sake of clarity, we first consider the case when $d=1$. Let N be a node of height h; then the corresponding playing board has $h+1$ squares. For even $h(>1)$, let N_2 (2 denotes the height of the node) be the node of the game position consisting of the middle 3 squares of the playing board of N. The playing board of N_2 is given by removing $(h/2-1)$ squares from each end of N; N_2 is the middle descendant of N (if N_2 is not N itself) of height 2. Similarly, if $h(>1)$ is odd, then let N_1 (of height 1) be the middle descendant of N, given by removing $(h-1)/2$ squares from each end of N. Then we have the following theorem.

Theorem 5.1. The Minimax value of a node N of height h (>2) in a G_1-game graph is always equal to the minimax value of its middle descendant of height 2 or 1, according to whether it is a MAX or MIN node:

$$M(N)=M(N_2)$$

if N is a MAX node (h is even);

$$M(N)=M(N_1)$$

if N is a MIN node (h is odd), where $M(\ )$ means "the minimax value of."

Proof. Suppose that $h>4$ is even; that is, N is a MAX node. The four shallowest levels of the game graph with N as the root are shown in Fig. 5.2. Let the symbols A, B_1, B_2, and so on represent both the nodes and the corresponding minimax values at the node positions. Since 1 and 0 are the only possible minimax values, the operations which maximize and minimize minimax values are actually the boolean operators OR and AND, respectively. Therefore, we have the following boolean

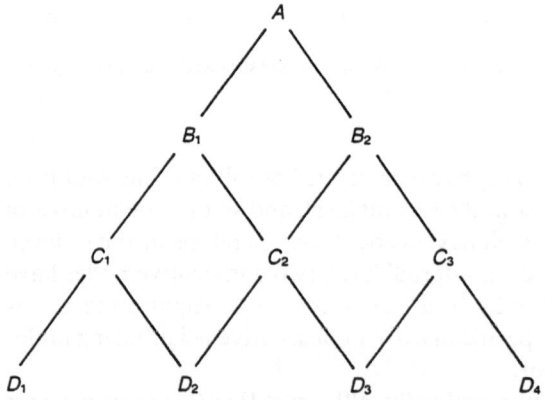

Fig. 5.2. The four shallowest levels of a G_1-game graph

equations:

$$A = B_1 + B_2$$
$$= C_1 C_2 + C_2 C_3$$
$$= (C_1 + C_3) C_2$$
$$= (D_1 + D_2 + D_3 + D_4)(D_2 + D_3)$$
$$= D_2 + D_3$$
$$= C_2 .$$

Now it is easy to show by induction that the minimax value A of the root is equal to the minimax value of the middle node of height 2. The case when N is a MIN node can be similarly proved by interchanging the operators, AND and OR. $\square$

Next we extend the simple derivation above to a general G_d-game. Let N be a node of height h; then the corresponding playing board has $hd + 1$ squares. For even $h (> 1)$, let N_2 be the middle descendant of N of height 2 ($N_2 = N$ if $h = 2$). The playing board of N_2 is given by removing $(h/2 - 1)d$ squares from each end of N. If $h (> 1)$ is odd, let N_1 be the middle descendant of N of height 1, of which the playing board is given by removing $d(h-1)/2$ squares from each end of N. Then we have the following general theorem.

Theorem 5.2. The minimax value of a node of height h (> 2) in a G_d-game graph is always equal to the minimax value of its middle descendant of height 2 or 1, according to whether the node is a MAX or MIN node:

$$M(N) = M(N_2)$$

if N is a MAX node (h is even);

$$M(N) = M(N_1)$$

if N is a MIN node (h is odd).

Proof. Suppose that $h > 4$ is even; that is, N is a MAX node. The four shallowest levels of the game graph, with N as the root, are shown in Fig. 5.3. Let the symbols A, B_1, and so on represent both the nodes and their corresponding minimax values. Using the boolean operators OR and AND, we have similar (to Theorem 5.1) but lengthier equations ($k = d + 1$):

$$A = B_1 + \ldots + B_k$$
$$= C_1 \ldots C_k + \ldots + C_k \ldots C_{2k-1}$$
$$= (C_1 \ldots C_{k-1} + \ldots + C_{k+1} \ldots C_{2k-1}) \cdot C_k ,$$

because C_k is the common successor of $B_1, \ldots, B_k$. Therefore, $A \leq C_k$. On the other hand, the k consecutive neighboring nodes $C_1, \ldots, C_k$ have the common successor $D_k; C_2, \ldots, C_{k+1}$ have the common successor D_{k+1}; and so on. Finally, $C_k, \ldots,$

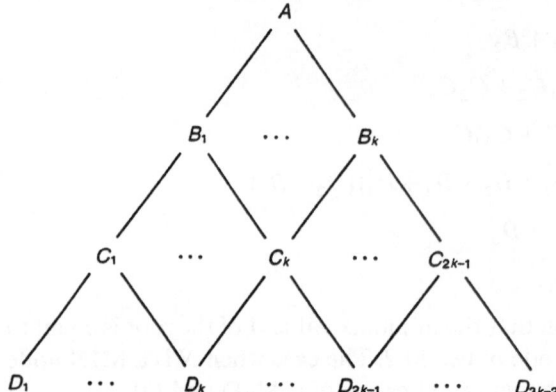

Fig. 5.3. The four shallowest levels of a G_d-game graph, where $k = d+1$

C_{2k-1} have the common successor D_{2k-1}. Since all nodes C_i are MAX nodes,

$$A = C_1 \ldots C_k + \ldots + C_k \ldots C_{2k-1}$$

$$\geq D_k + \ldots + D_{2k-1}$$

$$= C_k.$$

From the two inequalities above, we get $A = C_k$. Interchanging the operators, AND and OR, we get a similar result for the case when N is a MIN node. The theorem can be proved now by induction. $\square$

We have demonstrated that the minimax value of the root is always equal to the minimax value of the middle node at each level of the same type as the root. If the root is a MAX node, then the minimax value of the middle node at each MAX node level is always equal to the minimax value of the root. Similarly, if the root is a MIN node, then the minimax value of the middle node at each MIN node level is always equal to the minimax value of the root. Therefore, the minimax value of a node can always be reduced to the minimax value of the middle descendant of height 1 or 2, and the process of searching for minimax values can be drastically simplified.

Later we also demonstrate that, in a heuristic search of G_d-games with the one-counter as the static evaluation function, the above theorems lead to a very simple method of estimating minimax values using heuristic information.

6   Heuristic Information

In this chapter we formulate a concept of heuristic information. In a conventional heuristic game-tree search, heuristic information is a set of values returned by a static evaluation function at search-tip nodes. A static evaluation function is primarily used to estimate the strength of a node. However, the notion of node strength relies mostly on heuristic and intuition and usually does not have a precise formulation. Since the minimax value is the value most used to characterize a node in a game tree, the strength of a node is often seen as the minimax value of the node. But, a static evaluation function, such as the one-counter in P_b- and G_d-game models, usually does not directly estimate minimax values. In a P_b- or G_d-game, the minimax value of a node is correlated with the number of ones in the corresponding game position (that is, the value of the one-counter), but this number is definitely not a direct estimate of the minimax value. A transformation from this search information, the number of ones, to an estimate of the minimax value is necessary. Here we first study such search information, called heuristic information, as a mathematical object, and then consider node estimation based on heuristic information in the next chapter.

A value or a set of values of a static evaluation function represents certain information about the game being searched. For any other game (in the game model) that has this same information, the static evaluation function would return the same value or the same value set. Therefore, we use the set of all games that have the same information to characterize the information. A value or a set of values then represents the set of all games of the same value or the same value set (at a fixed node or a fixed node set). For theoretical reasons, we assume that such a game set is an event (Chap. 4) in the game model. Hence, we define a piece of heuristic information about a given game to be an event containing the given game.

Intuitively, more information describes a game more precisely. Therefore, we use the "containing relation" of events to compare heuristic information. That is, if I and J are two pieces of heuristic information (two events) about a game x and I is contained in J, then I is said to be more precise than J, as heuristic information about x. If I and J are any two pieces of heuristic information about x, we can accumulate them by the intersection operation, IJ, which is more precise than both I and J. Therefore, with our formulation of heuristic information, we can compare and accumulate heuristic information.

A heuristic search is defined as a function which returns (at each node) a piece of heuristic information about the game being searched. A conventional static evaluation function is a special heuristic search under this definition. In a search tree, we accumulate heuristic information (by means of the intersection operation) at all of the nodes in the search tree. The cumulative heuristic information is more precise than the heuristic information at any node of the search tree. And a heuristic search always returns more precise heuristic information at a larger set of nodes. This monotone property guarantees that we can search for more precise heuristic information by generating more nodes.

Accumulating heuristic information only at search-tip nodes, a conventional heuristic search does not necessarily return more precise information with a deeper search. This is a major problem with some pathology experiments, which fail to save information from further up in the tree. In several cases this information could have been used to break ties and possibly prevent pathology.

This chapter presents the precise formulation of heuristic information and the formal definition of heuristic search having the monotone property.

6.1   Examples: P_2- and G_1-Game Models

In this section, we first use P_2- and G_1-game models as examples to demonstrate how to develop a concept of heuristic information from the values returned by a static evaluation function. Using this concept, the comparison of heuristic information at different levels of the game tree becomes a straightforward process. The belief that a deeper search returns more precise information can be readily justified in the first example. However, for the second example this belief is not necessarily true. In order to get more information about a G_1-game, the relevant heuristic information at shallower levels should be retained. A precise formulation of heuristic information for general probabilistic models is introduced in Sect. (6.2), based on the discussion of these typical examples.

6.1.1 P_2-Game Models

P_2-game models have already been introduced in Chap. 2. Suppose that the one-counter, which counts the number of ones in each position, is the static evaluation function in our heuristic search. We now discuss the information reflected by the values returned by this one-counter and use the special game shown in Fig. 2.4 as our example.

For this special game, the one-counter returns 9 at the root, 5 at the left son, and 4 at the right son. The value 9 at the root means that, out of 16 cells in the corresponding board configuration, there are 9 cells covered with ones and 7, i.e., $(16-9)$ cells covered with zeros. Here we suppose that the information about the total number of cells is available. This description of the relevant board configuration can be treated as the heuristic information returned by the one-counter at the root, and the value 9 can be treated as a piece of heuristic information instead of as a purely numerical value. The value 5 at the left son represents another piece of heuristic information, namely, out of 8 cells in the left half of the board configuration, 5 cells are covered with ones, and 3 cells are covered with zeros. The heuristic information at the right son, represented by the value 4, is that 4 cells in the right half of the board configuration are covered with ones.

If we consider both the left son and the right son of the root and combine the two corresponding pieces of heuristic information, then we get a new piece of heuristic information represented by the pair (5, 4), in which 5 is the number of ones in the left half of the board configuration and 4 is the number of ones in the right half. From this piece of cumulative heuristic information, we can immediately determine that there are 9 ones in the whole board configuration; that is, the heuristic information at the root, represented by the value 9, is derivable from the cumulative information at both sons. However, this (cumulative) heuristic information, represented by the pair (5, 4), cannot be conversely derived from the heuristic information at the root (represented by 9). Let the number of ones in the right half of the board configuration be i. Then, from the value 9 at the root, the number of ones in the left half is $9 - i$. It is possible for i to vary from 1 to 8, because there are 8 cells in each half. However, there is no way to determine the exact value of i, given only the heuristic information at the root. Therefore, we conclude that the (cumulative) heuristic information at both sons is more precise than the heuristic information at the root.

Similarly, it is easy to show that the heuristic information accumulated at both the root and one of its sons is more precise than the information at only one son. For example, from the information at both the root and the left son, namely, 9 ones in the root and 5 ones in the left son, we can determine that $4 = (9 - 5)$ ones are in the right son. But we cannot determine the numbers (9 and 5) of ones in the root and in the left son from only the number (4) of ones in the right son. Moreover, the one-counter returns even more precise heuristic information at the nodes of the next level, this information being represented by the list (1, 4, 2, 2) of the numbers of ones in the corresponding parts of the board configuration. For a general P_2-game, a similar discussion shows that the cumulative heuristic information at a deeper level is always more precise. If we search the leaves, then we get the heuristic information completely characterizing the whole board configuration. We call such heuristic information the *complete information*.

The property that more precise information is returned at a deeper level coincides with the conventional belief that a static evaluation function returns more precise information at a deeper level. For P_2-games searched by the one-counter, this property stems from the fact that the number of cells at each node in the game tree is the sum of the number of cells at the two sons of the node. This property also holds for general P_b-game models.

In the above discussion, we use the description of the board configuration to represent the heuristic information returned by the one-counter. Next we consider another representation by equations of random variables. Let $X(=(X_1, \ldots, X_{16}))$ be the random vector representing a game in our P_2-game model, in which X_i is the value at the ith leaf from the left, and let the game given in Fig. 2.4 be $x(=(x_1, \ldots, x_{16}))$. With this game representation, the one-counter, counting the number of ones in each game position, finds the sum of corresponding components of X. For example, the value 9 at the root corresponds to the equation

A:
$$X_1 + \ldots + X_{16} = 9.$$

The values 5 at the left son and 4 at the right son correspond to the following equations, respectively:

B:
$$X_1 + \ldots + X_8 = 5,$$

C:
$$X_9 + \ldots + X_{16} = 4.$$

Therefore, each of these equations can be used to represent the corresponding heuristic information at each node.

In general, the value returned by the one-counter at a game position corresponds to an equation indicating that the sum of certain components of X is equal to this returned value. The cumulative heuristic information at a set of nodes, then, corresponds to the system of equations at all of the nodes in the set. For example, the cumulative heuristic information at both sons of the root corresponds to the following system of equations:

BC:
$$\begin{cases} X_1 + \ldots + X_8 = 5, \\ X_9 + \ldots + X_{16} = 4. \end{cases}$$

And the cumulative heuristic information at the next level corresponds to the following system of equations:

D:
$$\begin{cases} X_1 + X_2 + X_3 + X_4 = 1, \\ X_5 + X_6 + X_7 + X_8 = 4, \\ X_9 + X_{10} + X_{11} + X_{12} = 2, \\ X_{13} + X_{14} + X_{15} + X_{16} = 2. \end{cases}$$

Since the above equations refer to the special game x, the game x satisfies all such equations. In fact, each equation or system of equations essentially represents the set of games satisfying that equation or system of equations. We can thus use the symbol representing an equation or a system of equations to denote the corresponding game set. The equation A at the root represents, for example, the following set, also denoted by A:

$$A = \{ X \mid X_1 + \ldots + X_{16} = 9 \}.$$

Similarly, we get sets B, C, and D. The set of games corresponding to the system of equations BC is exactly the intersection of the sets B and C. The symbol BC denotes the system of equations B and C as well as the intersection of the sets B and C. Therefore, each piece of heuristic information returned by the one-counter can be represented by the corresponding game set, which contains the special searched game x. Furthermore, since the P_2-game space under consideration is discrete, these sets are measurable; that is, they are events in the probabilistic game space.

The comparison of different pieces of heuristic information, discussed above, is the same as the well-known inclusion relation of events. That is, two pieces of heuristic information are *equally precise* if and only if they both correspond to the same set of games; one piece of heuristic information is *properly more precise* than the other if the game set relative to the first piece is properly contained in the game set relative to the other piece. We use "more precise" for short to mean "equally precise or properly more precise." If any one of two such events is not contained in the other, then neither is more precise than the other, and they are not comparable.

The intersection of any two game sets, representing two pieces of heuristic information about x, is a piece of heuristic information more precise than either one alone. That is, accumulating different pieces of heuristic information is equivalent to intersecting the corresponding events. For example, the fact that the heuristic information at both sons is more precise than that at the root is shown by the following simple set relation:

$$BC \subset A.$$

We also notice that

$$BC = AB = AC = ABC.$$

This means that if we accumulate the different pieces of heuristic information in the ways shown above, we get the same piece of heuristic information as BC. The above relation is shown in Fig. 6.1, where the event A consists of $\binom{16}{9}$ ($=11440$) potential games, B consists of $\binom{8}{5}2^8$ ($=14336$) potential games, and C consists of $\binom{8}{4}2^8$ ($=17920$) potential games. These three events have $\binom{8}{5}\binom{8}{4}$ ($=3920$) potential games in common, of which $\binom{4}{1}\binom{4}{4}\binom{4}{2}\binom{4}{2}$ ($=144$) games form the set D.

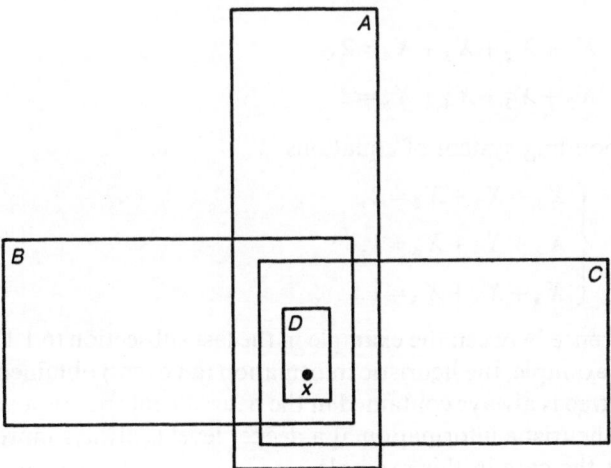

Fig. 6.1. Four pieces of heuristic information about a P_2-game x, where A is at the first level (the root), B and C are at the left and right nodes, respectively, of the second level, D is at the third level, and the heuristic information is always more precise at a deeper level, i.e., $A \supset BC \supset D$

The above representation of heuristic information by events in a probability space has already revealed two advantages. First, we can precisely accumulate different pieces of heuristic information by intersecting the corresponding events. Second, we can use the inclusion relationship between events to effectively compare heuristic information. Moreover, since the term "event" has a precise definition in probability theory, we can use this well-developed probability theory to study the heuristic information with the representation above. Therefore, in the next section, we formally define a piece of heuristic information in a probabilistic game model as an event in the model.

6.1.2 G_1-Game Models

The next example is the G_1-game model for the game graph in Fig. 2.2, the one-counter being the static evaluation function for the search of the given game. As in the previous example, the heuristic information returned by the one-counter is a description of a game position, and this description can be represented by either an equation or a system of equations.

Let $X(=(X_1\ X_2\ X_3\ X_4\ X_5))$ be the random vector representing a game in the model, and let $x(=(1\ 1\ 0\ 0\ 1))$ be the special game being searched. For this special game, the one-counter returns 3 at the root, and the corresponding piece of heuristic information is represented by the following equation A or event A, of games satisfying the equation:

A: $$X_1+X_2+X_3+X_4+X_5=3.$$

At both sons of the root, we get the following equations and the corresponding events B and C of games:

B: $$X_1+X_2+X_3+X_4=2,$$
C: $$X_2+X_3+X_4+X_5=2.$$

At the next level the corresponding system of equations is

D: $$\begin{cases} X_1+X_2+X_3=2, \\ X_2+X_3+X_4=1, \\ X_3+X_4+X_5=1. \end{cases}$$

There is an essential difference between the example in the last subsection (6.1.1) and this example. In the last example, the heuristic information (an event) obtained at a deeper level of the game tree is always contained in the heuristic information at a shallower level; that is, the heuristic information at a deeper level is always more precise. However, this is not the case in this example.

The above equation A is not derivable from equations B and C; that is, the event BC is not contained in the event A. For example, the game $y=(0\ 1\ 1\ 0\ 0)$ is in both B and C because both sons of y have two ones. However, since y itself has only two ones instead of three ones, y is not in A. Similarly, the game $z = (1\ 0\ 1\ 0\ 0)$ is in D, the heuristic information at the next deeper level, but z is not in BC nor in A.

The above relation is illustrated in Fig. 6.2, which indicates that the heuristic information collected at a deeper level is not necessarily more precise than that at a shallower level. Therefore, accumulating the information at search-tip nodes only, as in a conventional process, does not necessarily collect all available information. To get more precise information from searching more nodes, theoretically, we need to accumulate the heuristic information at all of the search nodes. Therefore, at level 2 the heuristic information should be ABC instead of BC, and at level 3 the heuristic information should be $ABCD$ instead of D alone. It is easy to determine that event $ABCD = \{x\}$; that is, we have already collected the complete information at all nodes down to level 3. In this new collecting process, the heuristic information always becomes more precise when the search is deepened, as is shown in the following séquence:

$$A \supset ABC \supset ABCD = \{x\}.$$

Later (Chap. 10) we come back to this example and show that it is possible to design, for this special model, a new process of accumulating heuristic information. This process is even simpler than the conventional search procedures.

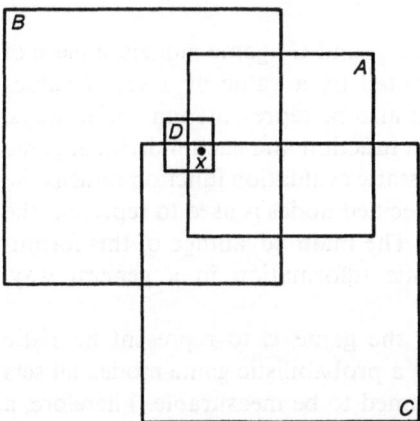

Fig. 6.2. Four pieces of heuristic information about a G_1-game x, where A is at the first level (the root), B and C are at the left and right nodes, respectively, of the second level, D is at the third level, and the heuristic information is not necessarily more precise at a deeper level

6.2 Formulation of Heuristic Information

In this section we give a formal definition of heuristic information in a probabilistic game model. This formulation, a generalization of the discussion of P_2- and G_1-game models in the previous section, is useful for the theoretical study of heuristic game-graph searches.

Hereafter in this chapter, the probabilistic game model under consideration, denoted by (Ω, F, P), is arbitrary but fixed. The space Ω is the space of all potential games; F is a Borel field of subsets of Ω, and P is a probability measure on F. Let T be the relevant game graph (or tree). Just as before, any element E (of F), where E is a set of games in the space with probability $P(E)$, is called a measurable set or an event. In many actual cases, such as P_2- and G_1-game models, since the whole space is finite or discrete and since each individual game has a probability (i.e., it is measurable), any set of games is measurable. That is, F consists of all subsets of games in the space Ω, and we need not doubt the measurability of any set.

The particular game given, denoted by x, is arbitrary but also fixed in this chapter. In this section we give a formulation of heuristic information about the game x. A process for finding heuristic information about x and a comparison between pieces of heuristic information are introduced in the subsequent sections.

Consider a conventional search process which searches the game graph T, the values at its leaves forming the given game x. A so-called static evaluation function is used to return a value at each search-tip node, this value being expected to reveal a certain property of the game x at the search-tip node. Any information implicitly or explicitly contained in values returned by the static evaluation function is usually called *heuristic information* about the game x. Therefore, the static evaluation function is treated in this book as a method of searching for heuristic information.

As we have already discussed (Sect. 6.1) for P_2- and G_1-game models, a piece of such heuristic information, originally represented by a value or a set of values returned by the static evaluation function, can also be represented in varied ways, depending on each individual static evaluation function and each individual game model. Here, the set of all games for which the static evaluation function returns the same value (or the same set of values) at the specified nodes is used to represent the corresponding piece of heuristic information. The main advantage of this formulation is that we can consider such heuristic information in a general way, independent of each individual model.

In general, we use game sets, containing the game x, to represent heuristic information about x. Since we are considering a probabilistic game model, all sets of games under consideration are further assumed to be measurable. Therefore, a piece of heuristic information is formally defined as follows.

Definition 6.1. A piece of heuristic information about the game x is defined as any measurable set of games containing x.

A measurable set in the game space is also called an event. Therefore, a piece of heuristic information about x is an event containing x, and vice versa. Thus, we use these two terms, "piece of heuristic information" and "event," interchangeably.

In a conventional search process with a static evaluation function, the value the latter returns at a given node corresponds to a piece of heuristic information about the searched game x. Further, this piece of heuristic information is just the set of all games for which the function returns this same value at this same given node. Here the static evaluation function should be a random function (i.e., measurable) in order to guarantee that this game set is an event. Therefore, we always assume the measurability of any static evaluation function under consideration. For example, the one-counter discussed earlier (Sect. 6.1) is measurable and returns a piece of heuristic information about the searched game at each node for P_2- and G_1-game models.

Consider the whole game space Ω. Since this set contains each game in the game model, Ω is a piece of heuristic information about each game. This heuristic information is already known prior to any search, and any other piece of heuristic information is always contained in this universal set (Ω). Therefore, Ω is the coarsest information about any game in the game space, and we call Ω the *trivial information*, that which is available without any search.

Consider the other extreme case, the singleton $\{x\}$, consisting of x itself, and assume that this one-element set is measurable. This heuristic information characterizes the game x completely and is called the *complete information* about x. In a conventional search, if all leaves of the game graph are searched and if the actual payoffs at all nodes are returned, then these payoffs form the components of the vector x, and the complete information about x is thus available. Any useful property depending only on x can be derived from this complete information. For example, the minimax value of any node in the game graph can be exactly determined from this searched vector x only. The conventional minimax procedure, searching the complete game tree, is a process of searching for the complete information about the searched game and for the minimax value of each node.

On the other hand, the alpha-beta procedure usually does not search all leaves of a game tree in finding the actual minimax value of the root. If we keep track of all payoffs at searched leaves, this set of values refers to a set (an event) of games which have the same payoffs at those searched leaves. This event is a piece of heuristic information about the searched game, supposedly returned by the alpha-beta procedure. If not all of the leaves are searched, this event generally consists of more than one game and is therefore not the complete information. No information about the unsearched leaves, for example, is available. Although not complete, information returned by the alpha-beta procedure is nonetheless sufficient for determining the minimax value of the root, which is the only goal of the procedure. Therefore, all games in this event have the same minimax value at the root. The alpha-beta procedure makes it usually unnecessary to search for the complete information about a game in order to determine the minimax value of the root. This shows that a process of collecting heuristic information can be designed according to the goal of the process. An efficient process may best be designed when the properties of the goal and the heuristic information relevant to the goal are fully analyzed.

6.3 Heuristic Search

A searching process usually collects only partial information about the searched game x. The heuristic information returned by the searching process is generally neither the complete information nor the trivial information (which contains no special information about x at all). It is often not feasible to search for the complete information because of the size of the game graph, and it is often easy enough to search for more information than the trivial information.

Since in our abstract model any event (a measurable set of games) containing the game x is treated as a piece of heuristic information about x, any process which can return heuristic information about each searched game is treated as a legitimate search in our abstract theory. A legitimate search, called a heuristic search, is formally defined as follows.

Definition 6.2. A *heuristic search* is a function which returns a piece of heuristic information about the game being searched at any node in the game graph T. All possible piece of heuristic information returned by the heuristic search at each fixed node form a partition of the whole game space.

Any piece of heuristic information (an event or a measurable set of games) returned by a heuristic search is also called a *search event* or heuristic search information. A heuristic search is a function of two variables, which maps each pair of a game in the game space and a node in the game graph to an event containing the game. Furthermore, for a fixed node, all of such events form a partition of the game space. Therefore, a heuristic search can be characterized by the corresponding partition at each node.

A conventional static evaluation function is a special heuristic search. If a static evaluation function returns a value v at a node N for the game x, then the heuristic information is the set of all games for which the static evaluation function also returns v at N. That is, each value at a node represents a piece of heuristic information at that node. The set of all possible values at a node then corresponds to a partition of the game space. For example, the one-counter discussed previously (Sect. 6.1) is a heuristic search for P_2- and G_1-game models.

Given a heuristic search and a set S of nodes in the game graph, we consider each search event returned by the heuristic search at each node of S. Since each search event at a node is a set of games containing the game x being searched, we get at all of the nodes in S the information indicating that the game x belongs to all of these search events; that is, x belongs to the intersection of the search events at nodes of S. Therefore, to accumulate heuristic information at nodes of S is to take into consideration the intersection of all pieces of heuristic information. We call the intersection of these search events (which is also an event containing x) the piece of heuristic search information or the search event returned by the heuristic search at the set S. A search event at a set of nodes is thus also a piece of cumulative heuristic information. All possible pieces of cumulative heuristic information at a fixed node set also form a partition of the game space.

In a conventional heuristic game-tree search, the heuristic information is usually collected at the set of all search-tip nodes; for example, a level of nodes in a game tree is often considered to be the set of search-tip nodes where the search stops. The heuristic information at such search-tip nodes is included in the list of values returned by the corresponding static evaluation function at these search-tip nodes. As we have demonstrated for P_2- and G_1-game models, such a list of values corresponds to an event, which is the intersection of all search events at those search-tip nodes.

With the intersection operation used to accumulate heuristic information, a heuristic search now becomes a process of searching at any set of nodes in the game graph for heuristic information about the game being searched. Such a set of nodes is called a set of *search nodes*, and the corresponding piece of heuristic information is called the *search event* or the *heuristic search information* at the set of search nodes.

6.4   Improved Visibility of a Heuristic Search

In a conventional heuristic game-tree search, it is usually assumed that a static evaluation function returns more precise estimations at deeper levels of the game tree. This property of a static evaluation function is the so-called *improved visibility*. However, this property has rarely been justified because there has been no precise formulation of heuristic information. In this section, we introduce the idea of visibility of a heuristic search in our abstract model; in this model heuristic information and heuristic search now have formal definitions.

Since a piece of heuristic information about the game x is an event, a game set containing x, the set relation "inclusion" can naturally be used to compare different pieces of heuristic information about x. Therefore, we define three relations as follows.

Definition 6.3. Let H and K be any two pieces of heuristic information about x.

(1) If H and K are equal, then H and K are *equally precise*.
(2) If H is properly contained in K, that is, if all games in H are also in K, and if there is at least one game in K which is not in H, then H is *properly more precise* than K.
(3) If H is not contained in K and K is not contained in H, then H and K are *non-comparable*.

We also use "more precise" as an abbreviation in place of "equally or properly more precise."

Two pieces of heuristic information about x are equally precise if they refer to the same event. In a conventional search, if two static evaluation functions return two different values at a node but these two values correspond to the same search event, then these two functions are equally precise at the node. For example, it is easy to see that the zero-counter (which counts the number of zeros in a game position) and the one-counter are equally precise at all nodes for P_2- and G_1-game models.

If one piece of heuristic information about x is properly contained in another piece, then the first piece gives more precise information about x. As discussed

earlier (Sect. 6.1) the two equations describing both sons of the root in the game tree of Fig. 2.4 define more constraints than the equation describing the root; therefore, the heuristic information at both sons is properly more precise than that at the root. In fact, it is easy to confirm that the one-counter always returns properly more precise heuristic information at a deeper level of nodes for general P_b-game models; that is, the conventional belief of improved visibility is true for the one-counter in P_b-game models.

For G_d-game models we reach, however, a different conclusion. For the G_1-game model in Fig. 2.2, we have shown (Sect. 6.1) that the heuristic information returned by the one-counter at both sons of the root is generally non-comparable to the heuristic information returned at the root. Each provides certain information (about the searched game) not available in the other. For general G_d-game models, the one-counter similarly does not always return more precise heuristic information at a deeper level of the game graph. This gives a counterexample to the concept of improved visibility mentioned above.

Next, we introduce the improved visibility of a heuristic search in a more general way. Consider any two pieces of heuristic information about x, H and K. If they are non-comparable, both pieces can be combined by considering their intersection, HK, as we did in accumulating heuristic information at a set of nodes in the game graph. Both H and K contribute to this cumulative information HK; that is, HK is properly more precise than both H and K as heuristic information about x. Neither H nor K can be discarded in this accumulating operation. However, suppose that one of them is more precise than the other; then the latter one can be discarded, and the first one is already the cumulative heuristic information whether or not we combine them. More than two pieces of heuristic information can be similarly combined by this intersection operation if necessary.

Consider a heuristic search in our model. A searching process generates part of the game graph (or game tree) from the root and searches the nodes thus generated with the given heuristic search. In a conventional searching process, the heuristic search is the given static evaluation function, which returns values at the search-tip nodes. Those values returned by the static evaluation function are then manipulated by a back-up process. In our abstract model, the heuristic search is a function which, as defined in the previous section, returns at each node a piece of heuristic information about the searched game.

Suppose a set of nodes is generated by the searching process. We consider the heuristic information which is collected at all the generated nodes. Such heuristic information is the search event at the set of the generated nodes. This event is the intersection of the pieces of heuristic information returned by the heuristic search at all the generated nodes. The difference between this model and a conventional model is that the heuristic information is conventionally collected only at the search-tip nodes, instead of at all of the generated nodes.

Since the search event returned in our searching process is the intersection of pieces of heuristic information at all of the generated nodes, more precise heuristic information is always returned if more nodes are generated in the searching process. Therefore, we get the following conclusion, formulating the improved visibility of a heuristic search.

Theorem 6.1. For a probabilistic game model, a heuristic search has an improved visibility; that is, a heuristic search returns more precise heuristic information if more nodes are searched.

Suppose that a heuristic search such as the one-counter in P_b-game models always returns more precise heuristic information at a deeper level of nodes. In a search down to a certain level of nodes, the heuristic information at the search-tip level then contains the heuristic information at any other shallower nodes. In this special case, heuristic information can be collected at only the search-tip nodes as in a conventional heuristic search.

In G_d-game models, however, the one-counter does not always return more precise heuristic information at a deeper level of nodes. For such cases, to obtain more precise heuristic information by generating more nodes, we need to collect heuristic information at both previously- and newly-generated nodes. Under this condition a totally new searching process may be designed, as we demonstrate later in this book for G_d-game models.

In this section, we have been concerned about the heuristic information in a heuristic search process. In the next chapter, we discuss ways to manipulate heuristic information, such as the back-up process in a conventional heuristic search.

7 Estimation and Decision Making

The primary purpose of a heuristic search is to gather heuristic information for making a move. In a conventional approach, the decision making is based on estimates of possible next moves. These estimates are derived by backing up original estimates at search-tip nodes returned by a static evaluation function. From this conventional approach, we develop in this chapter a general theory of decision making based on heuristic information.

Although a static evaluation function is often used to estimate the strength of nodes, the actual meaning of node strength is intuitive and still vague. However, the strength of a node is usually represented by a value; that is, a node strength is a function of games in the game model. Therefore, we assume in this book that any node strength is a random variable. Since a node strength is formulated as a random variable, estimating the strength of a node becomes a mathematical problem, namely, the problem of estimating a random variable.

Minimax values are most common examples of node strength. In fact, the minimax value of a node is the result of the game if both players play from the node and make moves by using actual minimax values (Sect. 2.6). In other words, the minimax value of a node is the game value of the node relative to minimax optimal strategies (Sect. 5.2). If both players use other strategies, then the strength of a node becomes the game value relative to those two strategies. In general, we use the game value relative to both players' strategies as the node strength if these two strategies are given. Such a game value is always a random variable. It is still unknown whether there are any other reasonable random variables for representing the strength of a node.

Since we formulate node strength as a random variable, estimating the strength of a node in the conventional approach now becomes a problem of estimating random variables. In this chapter, we first discuss random variable estimation based on heuristic information and then decision making using these estimates. We also derive a relation between estimates based on comparable pieces of heuristic information. This relation indicates that a more precise piece of heuristic information yields a more precise estimate.

We formulate a decision-making process from MAX's point of view; the purpose of the decision at a move is to maximize the payoffs at the end of the game. The corresponding theory of MIN's moves can be similarly developed. A decision model for a move is formulated as a set of random variables, each random variable representing a kind of strength of a possible next move. The ideal goal of the decision making is to maximize those random variables. For example, maximizing the minimax values of the next move is often used as a decision-making strategy in conventional approaches.

In a heuristic search, the decision is made by maximizing the estimates of the random variables in the decision model. The relation between decision qualities based on comparable pieces of heuristic information indicates that the decision is improved when it is based on more information.

Throughout this chapter, the probabilistic game model under consideration is fixed and is denoted by (Ω, P, F). T is the corresponding game graph, and S is the heuristic search in our discussion.

7.1 Random Variable Estimators

Based on the information returned by the heuristic search S, we consider estimates of an arbitrary (integrable) random variable, M.

Suppose that A is a set of nodes in the game graph T. Then for any given game x in our model, S returns at A a piece of heuristic information about x, denoted by I. Furthermore, for any game in I, S returns the same piece of heuristic information at A; all possible pieces of heuristic information returned by S at A form a partition of the whole game space Ω. Let F_A be the Borel field generated by this partition, a subfield of the original Borel field F in our model.

Since the random variable M is integrable, the conditional expectation of M relative to the Borel field F_A exists; this conditional expectation is used to define an estimator of M.

Definition 7.1. The *A-estimator* of M, relative to the heuristic search S, is the conditional expectation M_A of the random variable M with respect to F_A:

$$M_A = E_{F_A}(M).$$

The A-estimator M_A of M is also a random variable. For any game x, if I is the piece of heuristic information returned by S at A, then the value of M_A at x is the conditional expectation of M relative to I:

$$M_A(x) = E_I(M) = \frac{1}{P(I)} \int_I M \, dP.$$

That is, $M_A(x)$ is the average of M in I, where I is assumed to have a positive measure. This means that the A-estimator M_A estimates M by computing the average of M in each piece of heuristic information returned. Such an averaging method is a popular estimating method in computer science.

Let $\{I_i\}$ be the partition of the whole game space formed by all possible pieces of heuristic information returned by S at A. Since in each I_i the A-estimator M_A has a constant value $E_{I_i}(M)$, M_A is a linear combination of indicators of all I_i's:

$$M_A = \sum_i E_{I_i}(M) \cdot 1_{I_i}.$$

In the special case that $A = \varnothing$, the empty set, S always returns the whole game space Ω, and M gives the expected value of M, that is, the average of M over the whole game space:

$$M_\varnothing = E(M).$$

Since S does not distinguish between games in this case, the only reasonable estimate of M is the average of M.

Consider the set of all nodes in the game graph T; the set is also denoted by T. In this case, S always returns the complete information ($F_A = F$), and the value of M can always be exactly derived; therefore, the A-estimator of M is M itself:

$$M_T = E_F(M) = M.$$

Since M_T is itself M, then M_T is the best estimator of M.

If M is the minimax value of a node in a WIN-LOSS game where M has only two possible values – 1 (a forced win for MAX) and 0 (a forced win for MIN) – then the estimate of M, given I as a piece of heuristic information returned by the heuristic search, is the conditional probability that the node is a forced win for MAX (Tzeng and Purdom 1983):

$$E_I(M) = P(M = 1 | I).$$

For example, if n, returned by the one-counter, is the number of ones under a node N in a P_b-game or a G_d-game, then the estimate of the minimax value M of N is the conditional probability that $M = 1$, given n ones under the node:

$$P(M = 1 | L(N) = n),$$

where $L(N)$ denotes the number of ones under the node N.

7.2   Comparison of Estimators

The heuristic search S usually returns different pieces of heuristic information at different sets of nodes. In this section, we study the relation between estimators (of an integrable random variable M) based on different but comparable pieces of heuristic information.

Let A and B be two sets of nodes in the game graph T. To make the pieces of heuristic information returned by S at A and B comparable, we assume that one set is contained in the others, $A \subseteq B$.

For any game in the model, the heuristic search S returns more precise (i.e., equally precise or properly more precise) heuristic information at B than at A. Therefore, the partition of the whole space Ω formed by all possible pieces of heuristic information returned by S at B is finer than the partition formed by all possible pieces of heuristic information returned by S at A. That is, the corresponding Borel fields have the following relation:

$$F_A \subseteq F_B.$$

Since the A-estimator M_A and the B-estimator M_B are conditional expectations relative to F_A and F_B, respectively, from the property of conditional expectations (Theorem 4.2), we have the following relations:

$$M_A = E_{F_A}(M) = E_{F_A}(E_{F_B}(M)) = E_{F_A}(M_B).$$

Therefore, we have derived the following theorem.

Theorem 7.1. If $A \subseteq B$, then the A-estimator M_A of M is also the A-estimator of the B-estimator M_B:

$$M_A = (M_B)_A.$$

This theorem indicates that the value returned by the A-estimator M_A is the same as the average of the values returned by the B-estimator M_B; that is, M_B is more precise than M_A. This relation corresponds to the conventional belief that more precise information leads to a more precise estimation.

Consider the P_2-game model for the game tree in Fig. 2.4 as an example. Let A be the set consisting of the root only, B the set of the root and its two sons, and M the minimax value of the root. The heuristic search with the one-counter returns the number of ones in each node. Suppose that L denotes the number of ones in the root and that L_1 and L_2 denote the numbers of ones in the left son and the right son, respectively. Then the above relation between estimates becomes:

$$P(M=1 \mid L=n) = \sum_{n_1} P(L_1 = n_1, L_2 = n - n_1) P(M = 1 \mid L_1 = n_1, L_2 = n - n_1).$$

This relation between conditional probabilities can also be derived directly from the mathematical definitions of these conditional probabilities. Consider the general form in Theorem 7.1 again. In probability theory, the sequence $\{M_A, F_A; M_B, F_B\}$, with the property described in the theorem, is called a *martingale* (Chap. 4).

More generally, let $\{A_i\}$ $(1 \leq i \leq n)$ be a finite increasing sequence of sets of nodes in the game graph T: $A_1 \subseteq A_2 \subseteq \ldots \subseteq A_n$. Then the heuristic search S returns more precise heuristic information at A_i than at A_j if $i > j$ $(1 \leq j < i \leq n)$. For the corresponding estimators, M_{A_j} is also the A_j-estimator of M_{A_i} if $i > j$, and the sequence $\{M_{A_i}, F_{A_i}\}$ $(1 \leq i \leq n)$ forms a martingale. Therefore, M_{A_i} is a more precise estimator than M_{A_j} if $i > j$. In the next section, this property is used to demonstrate the improvement of the corresponding decision quality at a larger set of searched nodes.

7.3 Decision Making

The purpose of a heuristic search is to collect useful information for decision making. Consider the decision problem in which MAX is to make a move from a MAX node A. If the node A has n successors $A_1, \ldots, A_n$, then MAX has these n alternatives. This decision problem is denoted by $\{A_i\}$ $(1 \le i \le n)$.

The primary goal of MAX within the game is to maximize the final payoffs at the end, or, in the case of a WIN-LOSS game, to win the game. The criterion for solving the one-move decision problem formulated above should incorporate this primary goal of the whole game.

For the decision problem above, we first introduce a decision model in which each possible move is associated with a random variable. Formally stated, the goal of the move is to maximize these associated random variables. Such a random variable is commonly supposed to represent some kind of strength of the corresponding node.

Relative to the heuristic search S, we next consider, for a given decision model, a decision strategy and the corresponding decision quality. The relation between decision qualities, relative to the heuristic searches at different sets of nodes, is explicitly formulated; this relation indicates that the decision quality is improved if more nodes are searched.

7.3.1 Decision Models

Theoretically, at each move (of MAX) the corresponding complete game tree (or graph) can be generated and searched so that the minimax values of all possible next moves may be derived. Then, in each move MAX can use the strategy that always chooses a move with the largest minimax value.

If MIN uses a similar strategy, always making a move with the smallest minimax value, then all nodes chosen by MAX and MIN have the same minimax value. This value is ultimately the final payoff of the game. Both players' strategies are optimal.

In the game above, the criterion for each move (of MAX) is to maximize the minimax values of next moves; therefore, minimax values are chosen here to represent the node strength. This representation of strength is optimal if both players can search for the actual minimax value of each node. The corresponding strategies of MAX and MIN are the "minimax optimal strategies" (Chap. 5).

However, it has already been noted that, due to the size of game trees, searching for actual minimax values is usually not feasible. Most game-playing programs, in fact, use heuristic methods.

In a conventional heuristic approach, only part of the game tree is generated, and a "static evaluation function" is used to evaluate each search-tip node. Next, the values at the search-tip nodes are backed up by the well-known minimaxing procedure. Then MAX makes a move by choosing a node with the largest backed-up value.

One of the rationales behind this conventional approach is that the static evaluation function estimates the actual minimax values (or some kind of strength) of the search-tip nodes and, hence, the backed-up value at a node estimates the actual minimax value (or some kind of strength) of the node. This is the *face-value principle*: the values returned by the static evaluation function are treated as if they were actual minimax values.

However, the pathological phenomenon introduced earlier (Chap. 3) demonstrates the defect of this conventional approach; that is, the decision quality may deteriorate for a deeper search. In order to study the decision problem at a move, we use an abstract model in which the criterion of the move is formulated mathematically. The conventional pathological phenomenon can be explained in this new abstract model. The new model may also help in the exploration of new practical processes for decision making.

Formally, any kind of strength of a node is represented by a random variable. The criterion of the decision is to maximize these random variables which represent the strength of possible next moves. Therefore, we define a decision model for our decision problem as follows.

Definition 7.2. A *decision model* of the decision problem above is an association, denoted by $\{(A_i, M_i)\}$, $(1 \leq i \leq n)$, which binds each node A_i (a possible next move) with an integrable random variable M_i. M_i is called the *strength* of A_i, and the goal of the decision making is to maximize these M_i's.

Given a decision model, the heuristic search S gathers (at certain nodes) heuristic information which can then be used to estimate these M_i's. Then the decision is made by choosing a node with the maximum estimate.

If both players adopt the minimax optimal strategies by using actual minimax values after the current move, then the model in which M_i is the minimax value of A_i is the precisely accurate decision model. Since the minimax value of each A_i represents in this case the final payoff of playing the game from A_i, a move with the maximum estimate of minimax value represents a move with the maximum estimate of final payoff. That is, the minimax value of a node becomes the strength of the node if both players use the minimax optimal strategies from this node on.

If, however, after each possible move A_i, MAX uses a strategy $f \in R_{\text{MAX}}$, and MIN uses a strategy $g \in R_{\text{MIN}}$ (f and g may depend on A_i), the strength (M_i) of A_i becomes the game value of A_i, relative to f and g:

$$M_i(x) = M_{A_i}{}^{(f,\, g)}(x).$$

In this case, the expected final payoff of playing the game from A_i is the game value M_i.

If the behaviors of both players are not previously known, the choice of a decision model is still open. In this book, however, we assume that an arbitrary decision model is given, and we study a decision theory in this model.

Given a decision model, to estimate the strength of a move based on a heuristic search is to estimate the relevant random variable in the decision model, based on the heuristic information returned by the heuristic search. The problem of estimating random variables has been considered in the previous two sections. The corresponding decision quality is discussed in the next subsection.

7.3.2 Decision Qualities

Given a decision model $\{(A_i, M_i)\}$ $(1 \leq i \leq n)$ for the decision problem $\{A_i\}$ $(1 \leq i \leq n)$, we consider the decision making based on the heuristic search S. The goal of the decision in this model is to maximize the strength of the A_i's, the strength of each A_i being represented by a random variable in the model. And the role of the heuristic search S is to gather heuristic information for estimating those random variables.

Let N be a set of nodes in the game graph T. Consider both the heuristic information returned by S at N and the N-estimators of the M_i's. Suppose that I is the piece of heuristic information returned by S (at N) for a given game x. Based on I, the estimate of the strength of A_i is

$$E_I(M_i), \quad (1 \leq i \leq n).$$

The corresponding decision-making task is to choose a node with the maximum estimate $\max_i E_I(M_i)$. If more than one node has the maximum estimate, then a node can be chosen randomly from those nodes with the maximum estimate, each node being an equally likely choice. Such a decision-making process is called *N-decision-making*.

Notice that any other distribution of nodes with the maximum estimate is also acceptable. For simplicity, we are using a uniform distribution among those nodes.

In the following, a new random variable, denoted by M, is introduced to represent the strength of N-decision-making. If A_i is the only node with the maximum estimate for any given game x, then we define:

$$M(x) = M_i(x).$$

M_i depends only on I, which is a piece of heuristic information containing x. If more than one node has the maximum estimate, then we define:

$$M(x) = \frac{1}{m} \sum_i M_i(x),$$

where each random variable M_i has the maximum estimate in I and m is the number of random variables. We call $M(x)$ the *N-decision random variable*.

Consider the average (mean) of the N-decision random variable M. Let $\{I_j\}$ be the partition of the whole game space generated by all possible pieces of heuristic information returned by S at N.

Then

$$E(M) = \sum_j P(I_j) E_{I_j}(M).$$

Since S returns I_j for all games in I_j, M has the same value for all x in I_j:

$$M(x) = M_i(x), \quad \text{or} \quad M(x) = \frac{1}{m} \sum_i M_i(x),$$

where M_i has the maximum estimate $\max_i E_{I_j}(M_i)$ in I_j. Therefore,

(Q) $$E(M) = \sum_j P(I_j) \max_i E_{I_j}(M_i).$$

Note that the right side of the equation above is independent of the N-decision random variable.

At the set N of nodes, the heuristic search S cannot distinguish between games in each piece of heuristic information I_j. Thus, the N-decision random variable, which denotes the strength of the node chosen by N-decision-making, has the maximum average in each I_j and, therefore, has the maximum average in the whole game space. This means that N-decision-making is optimal in terms of the average of the strength of the chosen node. Thus we have the following definition.

Definition 7.3. The decision quality of N-decision-making, denoted by $Q(N)$, is defined to be the average of the corresponding N-decision random variable, which is represented by the equation (Q) above:

$$Q(N) = E(M) = \sum_j P(I_j) \max_i E_{I_j}(M_i).$$

If the decision is based on the complete information, by which each M_i can be exactly observed, then the node chosen always has the maximum strength, and the decision making is perfect; that is, the goal of the decision model is achieved. However, the heuristic search S usually returns only partial information. The corresponding decision making is, therefore, not perfect. In the following paragraphs, we consider the relation between decision qualities of different ways of making decisions.

Let N_1, N_2 be two sets of nodes in the game graph T such that $N_1 \subseteq N_2$. Consider the relations between decision making relative to N_1 and to N_2.

Let $\{I_j\}$ be the partition of the whole game space formed by all possible pieces of heuristic information returned by S at N_1. Then all possible pieces of heuristic information returned by S at N_2 form a finer partition, denoted by $\{I_{jk}\}$, such that for each j:

$$I_j = \bigcup_k I_{jk}.$$

Notice that the index k depends on the index j.

For each game in an I_j, the N_1-decision-making chooses a node from $\{A_i\}$ with the following maximum estimate:

$$\max_i E_{I_j}(M_i).$$

And for each game in an I_{jk}, the N_2-decision-making chooses a node from $\{A_i\}$ with the following maximum estimate:

$$\max_i E_{I_{jk}}(M_i).$$

For each i and j, we have:

$$E_{I_j}(M_i) = \sum_k \frac{P(I_{jk})}{P(I_j)} E_{I_{jk}}(M_i)$$

$$\leq \sum_k \frac{P(I_{jk})}{P(I_j)} \max_i E_{I_{jk}}(M_i).$$

Thus, the decision qualities $Q(N_1)$ and $Q(\dot{N}_2)$ satisfy the following relations:

$$Q(N_1) = \sum_j P(I_j) \max_i E_{I_j}(M_i)$$

$$\leq \sum_j P(I_j) \sum_k \frac{P(I_{jk})}{P(I_j)} \max_i E_{I_{jk}}(M_i)$$

$$= \sum_{jk} P(I_{jk}) \max_i E_{jk}(M_i)$$

$$= Q(N_2).$$

The above relation is summarized in the following theorem.

Theorem 7.2. If N_1 and N_2 are two sets of nodes, then in a decision model $\{(A_i, M_i)\}$ the decision quality has the following monotone property:

$$Q(N_1) \leq Q(N_2) \quad \text{if} \quad N_1 \subseteq N_2.$$

The above theorem indicates that, in each decision model, the decision quality can be improved if more nodes in the game graph are searched; therefore, the pathological phenomenon in the conventional heuristic search process does not exist in our abstract decision model.

The essential difference between decision making in our abstract decision models and conventional decision making is that the actual process of estimating node strength in our model is implicit in each A-estimator whereas the minimaxing process is always explicitly used in conventional decision making. In our abstract models, the actual computation of the estimate of node strength is a purely mathematical problem. Therefore, this computation depends on each individual case.

In the next chapters, some actual games are discussed and various processes for estimating nodes are derived. The pathological phenomenon sometimes arises in the minimaxing process. Therefore, the conventional minimaxing process is not suitable for every game.

8 Independence and Product-Propagation Rules

Both the game models and the heuristic information introduced in the previous chapters are purely theoretical. Although estimates of the strength of nodes, given a piece of heuristic information, exist and have the monotone property (Theorem 7.1), no practical process of finding such estimates can be generally derived for all game models. Since these estimates have been formulated into mathematical terms, developing any practical process of determining them becomes a mathematical problem. Hence an actual process can only be created for each particular case.

This chapter introduces a special product model, which is a product of its components. Each component itself is also a probabilistic game model, and the strength of a node in each component is a local property depending only on that component. Specifically, for a product model of WIN-LOSS games, the product-propagation rules introduced by Pearl (1981, 1983) can be explicitly derived as a back-up process for estimating minimax values.

8.1 Product Models

Let (Ω, P, F) be a probabilistic game model for a game graph T, and let S be a corresponding heuristic search. In the following discussion, we demonstrate step by step how the probabilistic game model can be represented as a product of its components.

First consider the game graph T. Let A be the root of T and $A_1, \ldots, A_n$ the successors of A. Suppose that each A_i $(1 \leq i \leq n)$ does not have any predecessor but A, and that any two of these successors do not have a common descendant. For each $i(1 \leq i \leq n)$, let T_i be the subgraph consisting of A_i and all its descendants. Each T_i itself is also a game graph with A_i as the root, and these T_i's $(1 \leq i \leq n)$ are mutually disjoint. Such a T is called a product of its subgraphs, denoted by

$$T = \prod_{i=1}^{n} T_i .$$

If T is a tree, as in P_b-game models, then T is always a product of its subtrees. However, the game graphs for G_d-game models are not products of their subgraphs because two sibling nodes always have a common successor in such graphs and the subgraphs starting from the successors of the root are not disjoint.

Now suppose that T is a product of its subgraphs $T_i (1 \leq i \leq n)$. For each game X in the model, let X_i be the ith component of X at the leaves of T_i. Then X can be represented as:

$$X = (X_1, X_2, \ldots, X_n).$$

Each X_i is a subgame of X and has T_i as its game graph. Let Ω_i be the set of X_i's for all games in Ω.

Suppose that for each $i (1 \leq i \leq n)$ there is a probabilistic space (Ω_i, P_i, F_i), which is a probabilistic game model for the game graph T_i. Now we can define a product game model as follows.

Definition 8.1. Let (Ω, P, F) be a probabilistic game model for T. If the game graph T is a product of its subgraphs, and if the probability space (Ω, P, F) is a product space of (Ω_i, P_i, F_i), $1 \leq i \leq n$, then the game model (Ω, P, F) is called a *product game model* of (Ω_i, P_i, F_i), $1 \leq i \leq n$, and each (Ω_i, P_i, F_i) is called a component of the product model.

Hereafter, in this chapter, we generally assume that (Ω, P, F) is a product model of its components, as illustrated in Fig. 8.1. Note that we have the following product relations:

(*) $$\Omega = \prod_{i=1}^{n} \Omega_i, \quad F = \prod_{i=1}^{n} F_i, \quad \text{and} \quad P = \prod_{i=1}^{n} P_i.$$

For example, (Ω, P, F) can be a P_b-game model. Here $n = b$, and all subtrees T_i are isomorphic; that is, they all have the same shape and the same number of nodes. Therefore, all (Ω_i, P_i, F_i) are identical to the probability space of the P_b-game model for each T_i. Since all probability measures on Ω and the Ω_i's are induced by the

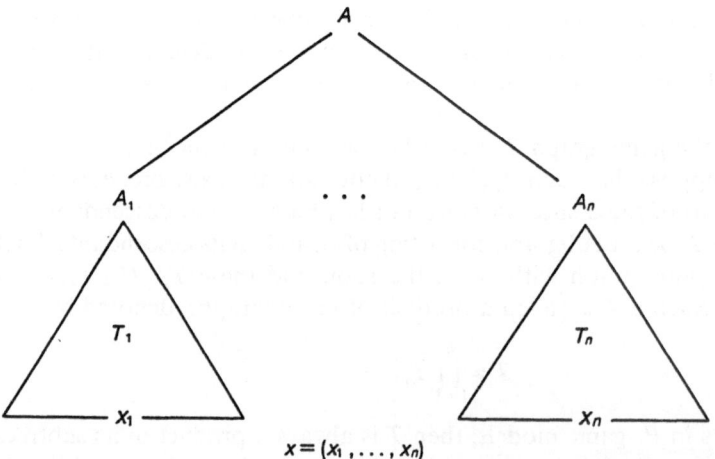

Fig. 8.1. A product game graph T, where A is the root, $A_1, \ldots,$ and A_n are successors of A, each T_i is the subgraph starting from A_i, each X_i is the game in T_i $(1 \leq i \leq n)$, and X is the whole game in T

identical, independent distributions of the values at the leaves of T, we have the product relations described above.

This discussion can be repeatedly extended to each component of a P_b-game model. Therefore, a P_b-game model is a hierarchy of product models; that is, the game model of all subgames at each internal node is a product model. Furthermore, these game models at all nodes of each level are identical.

Consider again the product assumptions in equations (*) above. Any random variable M on Ω_i induces a random variable $\bar{M}$ on Ω in a natural way:

$$\bar{M}(X_1, X_2, \ldots, X_n) = M(X_i).$$

Note that, if M is integrable, then $\bar{M}$ is also integrable, and both $\bar{M}$ and M have the same mean value. Therefore, we use the same notation, M, to denote both M and $\bar{M}$. For example, either the minimax value or a general game value of a node in T_i, being a random variable on Ω_i, is also treated as a random variable on Ω.

Let M_i be a random variable on Ω_i for each i. Then, from a property of product probability spaces (Sect. 4.2), $\{M_i\}$ ($1 \le i \le n$) becomes a set of independent random variables on the product space. Therefore, minimax values or general game values in different components are independent random variables in the product space. This independence is sometimes crucial in developing a practical process for estimating the strength of a node, as we demonstrate in the next section.

8.2 Product Heuristic Information and Product Heuristic Searches

Let $x = (x_1, \ldots, x_n)$ be an arbitrary game in the product game model (Ω, P, F). First, for each $i(1 \le i \le n)$, we introduce the concept of local heuristic information about the game x, relative to the ith component.

Consider the subgame x_i in the ith component, (Ω_i, P_i, F_i), of our game model. Suppose that J_i is a piece of heuristic information in the component about x_i; that is, J_i is an element of F_i containing x_i. Let I_i be the unique element of F defined by:

$$I_i = \{(X_1, \ldots, X_n) : X_i \in J_i \quad \text{and} \quad X_j \in \Omega_j \quad \text{if} \quad j \neq i\},$$

which, being a product of J_i and the Ω_j's ($j \neq i$), is an element of F. Since a random game X is in I_i if and only if its ith component, X_i, is in J_i, I_i is a piece of heuristic information about the game x, having information only about the ith component of x; it is called a piece of *local heuristic information* about x, relative to the ith component. J_i is called the *localization* of I_i.

Suppose that, for each $i(1 \le i \le n)$, I_i is a piece of local heuristic information about x, relative to the ith component, and J_i is the localization of I_i. Consider the product set

$$I = J_1 \times \ldots \times J_n,$$

which is a piece of heuristic information about x. In fact, I is the intersection of all I_i's:

$$I = I_1 I_2 \ldots I_n;$$

that is, a game X is in I if and only if it is contained in the local heuristic information I_i for each i. Such I is called a piece of *product heuristic information,* and I_i is called the local heuristic information of I, relative to the ith component.

Now let S_i be a heuristic search in the ith component (Ω_i, P_i, F_i) containing the subgame X_i. S_i returns at any node of the subgraph T_i a piece of heuristic information, say J_i, in the component. Let I_i be the piece of local heuristic information about x (relative to the ith component) with J_i as its localization. Since I_i is uniquely determined by J_i, we say that S_i returns I_i when the product game model, (Ω, P, F), is under consideration. Thus, we can also treat S_i as a heuristic search in the product game model. However, S_i is defined only at the nodes of T_i. At any node outside the subgraph T_i, we may let S_i return the whole game space Ω; that is, S_i may or may not return any special information outside T_i. Such a heuristic search is called a *local heuristic search,* relative to the ith component.

Suppose that, for each $i(1 \leq i \leq n)$, S_i is a heuristic search in the ith component (Ω_i, P_i, F_i). We also use the notation S_i to denote the local heuristic search in the product model, relative to the ith component. Now we define a product heuristic search as follows.

Definition 8.2. A heuristic search S in the product game model (Ω, P, F) is said to be a product heuristic search of S_i's, denoted by:

$$S = \prod_{i=1}^{n} S_i,$$

and S_i is said to be the ith component of S if at any node of the subgraph $T_i(1 \leq i \leq n)$ S is the local heuristic search S_i, relative to the ith component.

We do not have any restrictions on the heuristic information returned by a product heuristic search at the root of T; therefore, a product heuristic search is not uniquely determined by the local heuristic searches S_i. A product heuristic search S may return at the root of T some information about the components of the searched game.

A product heuristic search returns, at a set of nodes which are only in the subgraph T_i, a piece of local heuristic information (relative to the ith component) whose localization is returned by the ith component of the product search. Let N be a set of nodes in the game graph T, the root not being contained in N. Let N_i be the set of nodes of T_i which are in N, that is, the intersection of N and T_i. For each $i(1 \leq i \leq n)$, let I_i be the local heuristic information (relative to the ith component), about any given game, returned by S at N_i, and let J_i be the localization of I_i. Let I be the heuristic information returned by S at N. Then we have the following relations:

$$I = I_1 I_2 \ldots I_n = J_1 \times J_2 \times \ldots \times J_n.$$

That is, I is the intersection of the I_i's; I is also the product of the J_i's.

For example, consider again a P_b-game model, which is a product model. For any given game in the model, the number of ones under a non-root node is a piece of heuristic information referring only to the subgame under that node; therefore, the number of ones under a node is local heuristic information. That is, the heuristic search with the one-counter is a product heuristic search, which searches for local heuristic information in each subtree. However, at the root, the heuristic information, the number of ones under the root, is a piece of information about the successors of the root.

Since each component of a P_b-game model, treated as a product model, is also a P_b-game model, the discussion above can be extended repeatedly to all components of the P_b-game model.

8.3 Product-Propagation Rules

In this section we consider the estimation of random variables on a product game model (Ω, P, F) for a game graph T. Suppose that T is a product of its subgraphs T_i rooted at the sons of the root of T:

$$T = \prod_{i=1}^{n} T_i,$$

and that (Ω, P, F) is a product model:

$$(\Omega, P, F) = \prod_{i=1}^{N} (\Omega_i, P_i, F_i),$$

where each (Ω_i, P_i, F_i) is a game model for T_i.

Let M_i be an arbitrary random variable on the ith component, Ω_i. M_i can be extended to a random variable on the whole space in the following way:

$$X = (X_1, \ldots, X_n) \rightarrow M_i(X_i).$$

We also use the same symbol, M_i, to denote this extended random variable. For each $i(1 \leq i \leq n)$, let J_i be an element of F_i, and let I_i be the local heuristic information with J_i as the localization. Let I be the product heuristic information of all J_i's:

$$I = J_1 \times J_2 \times \ldots \times J_n = I_1 I_2 \ldots I_n.$$

If the random variable M_i is also integrable, then from Fubini's Theorem the estimate of M_i, given I, is

$$E_I(M_i) = \frac{1}{P(J_1 \times \ldots \times J_n)} \int_{J_1 \times \ldots \times J_n} M_i dP$$

$$= \frac{1}{P(J_1) \ldots P(J_n)} \int_{J_1} \ldots \int_{J_n} M_i dP_1 \ldots dP_n$$

$$= \frac{1}{P(J_i)} \int_{J_i} M_i dP_i = E_{I_i}(M_i) = E_{J_i}(M_i),$$

where $E_{J_i}(M_i)$ is the estimate of M_i as a random variable in the submodel, given J_i. Therefore, we have proved the following theorem.

Theorem 8.1. Let M_i be a random variable on $\Omega_i (1 \leq i \leq n)$. Given a piece of product heuristic information, the estimate of M_i depends only on the corresponding i^{th} local heuristic information.

Therefore, in order to estimate a random variable on any component of a product model, it is sufficient to search for local heuristic information relative to that component. And further, it is sufficient for a product heuristic search to search only the corresponding subgraph of the game tree.

Next we consider global minimax values on a product game model. First we assume, in addition to the production assumptions (*), that all games in Ω are WIN-LOSS games. At a leaf, let the value 1 represent a WIN and 0 a LOSS, relative to MAX. Then the minimax value of a node has only two possibilities, 1 and 0, denoting a forced win and a forced loss, respectively.

Let A be the root of T, and let A_i be the root of $T_i (1 \leq i \leq n)$. Suppose that M and M_i are the minimax values of A and A_i, respectively. Furthermore, the M_i's $(1 \leq i \leq n)$ are independent random variables (Sect. 8.1). We then have the following relations:

(**)
$$M = \min_i M_i = \prod_i M_i$$

if A is a MIN node, and

(***)
$$M = \max_i M_i = 1 - \prod_i (1 - M_i)$$

if A is a MAX node, where $\prod$ is the arithmetic product notation. The second relation is equivalent to:

$$1 - M = \prod_i (1 - M_i).$$

That is, A is a forced win for MIN if and only if all the A_i's are forced wins for MIN.

Given a piece of product heuristic information:

$$I = J_1 \times \ldots \times J_n = I_1 \ldots I_n,$$

J_i being the localization of the ith local heuristic information I_i, consider the estimate of M. There are two possible cases:

Case 1: A is a MIN node. Based on the equation (**), the estimate of M, given I, is derived as follows:

$$E_I(M) = \frac{1}{P(J_1 \times \ldots \times J_n)} \int_{J_1 \times \ldots \times J_n} \prod_i M_i dP_1 \ldots dP_n$$

$$= \prod_i \frac{1}{P(J_i)} \int_{J_i} M_i dP_i$$

$$= \prod_i E_{J_i}(M_i)$$

$$= \prod_i E_{I_i}(M_i).$$

Therefore, we have

(PP-1) $$E_I(M) = \prod_i E_{I_i}(M_i).$$

That is, the estimate at the root A, based on product heuristic information, is the product of the estimates at successors A_i, each based on the corresponding local heuristic information. Here the estimate at a node, based on a piece of heuristic information, is the conditional probability of a forced win (for MAX), given the heuristic information.

Case 2: A is a MAX node. Based on the equation (***), we have this corresponding derivation:

$$E_I(M) = E_I\left(1 - \prod_i (1 - M_i)\right)$$

$$= 1 - E_I\left(\prod_i (1 - M_i)\right)$$

$$= 1 - \frac{1}{P(J_1 \times \ldots \times J_n)} \int_{J_1 \times \ldots \times J_n} \prod_i (1 - M_i) dP_1 \ldots dP_n$$

$$= 1 - \prod_i \frac{1}{P(J_i)} \int_{J_i} (1 - M_i) dP_i$$

$$= 1 - \prod_i E_{J_i}(1 - M_i)$$

$$= 1 - \prod_i (1 - E_{J_i}(M_i))$$

$$= 1 - \prod_i (1 - E_{I_i}(M_i)).$$

Consequently, we have

(PP-2) $$E_I(M) = 1 - \prod_i (1 - E_{I_i}(M_i)).$$

Thus the estimate at the root A, based on a piece of product heuristic information, is also a function of estimates at all successors, each based on the corresponding local heuristic information. A slight modification leads to:

(PP-2)′ $$1 - E_I(M) = \prod_i (1 - E_{I_i}(M_i)).$$

This means that the conditional probability of a forced loss (for MAX) at A is the product of all conditional probabilities of a forced loss at each successor. We summarize the results above as the following theorem.

Theorem 8.2. Given a piece of product heuristic information, the estimates of the minimax values of the root and its sons satisfy (PP-1) or (PP-2) according to whether the root is a MIN or a MAX node.

The rules (PP-1) and (PP-2) are the so-called *product-propagation rules* (PP rules). These rules can be applied repeatedly if each component (Ω_i, P_i, F_i) is also a product model. (PP-1) is used to estimate minimax values at MIN nodes, and (PP-2) is used to estimate minimax values at MAX nodes.

Consider P_b-game models for example. Since all P_b-game models are product models, and since the one-counter is a product heuristic search, the PP rules can be repeatedly applied in estimating minimax values. Therefore, using the PP rules as the back up process, the estimate at the root can be derived by backing up the estimates at search-frontier nodes. The estimate at a search node (that is, the conditional probability of a forced win for MAX, given the number of ones under the node) is studied in the next chapter.

9 Estimation of Minimax Values in P_b-Game Models

A P_b-game model (Sect. 5.3) is characterized by its complete, uniform game tree with branching factor b and by the distribution of the payoffs at terminal nodes. The value 1 or 0 is independently assigned to all leaves with a probability of p or $1-p$, respectively, where 1 denotes a win for MAX and 0 a win for MIN.

In this chapter, we discuss the estimation of minimax values in P_b-game models, based on the one-counter as the heuristic search, which returns at any node the number of ones at the leaves under that node (Tzeng and Purdom 1986). Since the one-counter is a product heuristic search and each P_b-game model is a product model of other P_b-game models, the product-propagation rules derived in the last chapter can be applied repeatedly down to the search-tip nodes. The subject of this chapter is the estimation at a search-tip node, the only problem remaining.

9.1 More About Probabilities on P_b-Game Trees

Let T be the game tree of a P_b-game; hence, T is a complete, uniform tree with the branching factor $b\ (> 1)$. Each terminal node of T independently assumes the value 1 or 0 with a probability of p or $1-p$ $(0 \leq p \leq 1)$, respectively. For this chapter, assume that p is fixed.

Let H be the function *height* defined as follows.

$$H(A)=0 \qquad \text{if } A \text{ is a leaf, or}$$
$$H(A)=H(B)+1 \quad \text{if } B \text{ is a son of } A.$$

Since T is complete, the function H is well defined. Note that the number of leaves under a node A is $b^{H(A)}$. Let $L(A)$ denote the number of ones at the leaves under A. Then, $b^{H(A)} - L(A)$ is the number of zeros under A, and

$$0 \leq L(A) \leq b^{H(A)}.$$

Since the probability that a leaf has the value 1 is p and the values at leaves are independent, the distribution of $L(A)$ (the probability that $L(A) = l$) is

$$P(L(A)=l)=\binom{b^h}{l}p^l(1-p)^{b^h-1},$$

where $H(A) = h$, $(0 \leq l \leq b^h)$. Let $A_1, \ldots, A_b$ be the successors of A; then,

$$L(A) = \sum_{i=1}^{b} L(A_i).$$

Let the l_i's $(1 \leq i \leq b)$ be integers such that $0 \leq l_i \leq b^{h-1}$ and $\Sigma l_i = l$; then, we have the following conditional probability:

$$P(L(A_1) = l_1, \ldots, L(A_b) = l_b \mid L(A) = l)$$

$$= \frac{P(L(A_1) = l_1, \ldots, L(A_b) = l_b)}{P(L(A) = l)}$$

$$= \frac{P(L(A_1) = l_1) \ldots P(L(A_b) = l_b)}{P(L(A) = l)}$$

$$= \frac{\binom{b^{h-1}}{l_1} \cdots \binom{b^{h-1}}{l_b}}{\binom{b^h}{l}}.$$

Notice that the distribution of the number of ones under a node is a binomial distribution.

Let A be the root of the game tree T, and let $A_1, \ldots, A_b$ be the successors of A. Suppose that T_i is the subtree of T, with A_i as the root for each i $(1 \leq i \leq n)$. For each i, all subgames from A_i form a P_b-game model for the subtree T_i. Since all the T_i's $(1 \leq i \leq n)$ have the same tree structure, the corresponding P_b-game models are identical, and the original P_b-game model for the tree T is the product of these b identical P_b-game models for the b subtrees T_i.

The above discussion can be extended to each component; that is, the P_b-game model for each T_i is also a product model of its own components. Therefore, given any fixed node N of T, the P_b-game model consisting of all the subgames from N is a product model. Furthermore, all the P_b-game models at nodes of the same level are identical.

Let M_i be the minimax value of $A_i (1 \leq i \leq n)$. Since all of the models for the various A_i's $(1 \leq i \leq n)$ are identical, the M_i's are independent and identically distributed. This assertion is also true for all nodes at the same level; that is, the minimax values of all nodes at the same level are independent and identically distributed (i.i.d.).

9.2   The Conditional Probability of a Forced Win, $p(h, l)$

Consider the one-counter, L, as the heuristic search in our P_b-game model. If N is a node of the game tree T, $L(N)$ is the number of ones at the leaves under N for the game assigned to the tree T. The value $L(N)$ depends only on the subtree of T

rooting at N and is a piece of local heuristic information in the model, relative to the subtree. Therefore, L is a product heuristic search, as defined in the last chapter.

Since the P_b-game model at each node is a product model of WIN-LOSS games and the one-counter is a product heuristic search, the product-propagation rules (Chap. 8) are repeatedly applicable in estimating minimax values. Therefore, the problem of estimating the minimax value of the root of a P_b-game tree eventually becomes the problem of estimating minimax values of the search-tip nodes, based on the number of ones under each search-tip node.

Suppose that the game tree T is searched down to a particular level. Consider the estimation of the minimax value of the root, based on the heuristic search information. The estimation process consists of the following two steps: first, to estimate the minimax value of each node at the search-frontier level, based on the number of ones under each node, and second, to back up those estimates at the search-frontier level by repeatedly applying the product-propagation rules. At each node, the backed-up value is the estimate of the minimax value of that node. Finally, the backed-up value at the root is the estimate of the minimax value of the root.

Now consider the estimation of the minimax value of a search-tip node, based on the number of ones under the node. Since the minimax values of nodes at the same level are i.i.d. (Sect. 9.1), the estimate at each node depends on both the height of the node and the number of ones under the node. The order of the node at the level is irrelevant to this estimate.

Let h be the height of a node, and let l be the heuristic information returned at the node by the one-counter. Note that l, being the number of ones under a node of height h, satisfies $0 \le l \le b_h$. Since the estimate of the minimax value of the node (given l as the heuristic information) depends only on h and l, we use the symbol

$$p(h, l)$$

to denote such an estimate, the conditional probability of a forced win at any node of height h, given l ones under the node.

The actual values of $p(h, l)$ can be calculated by using the recurrence relations derived below. First, for $h=0$ it is trivial that:

(A) $p(0, 1)=1, \quad \text{and} \quad p(0, 0)=0$.

For $h=1$, there are two different cases to consider. If the nodes of height 1 are MAX nodes, then

$$p(1, 0)=0, \quad \text{and} \quad p(1, l)=1 \quad (l>0).$$

If they are MIN nodes, then

$$p(1, b)=1 \quad \text{and} \quad p(1, l)=0 \quad (l<b).$$

Now let $h(>1)$ be a positive integer, and let l be an integer such that $0 \le l \le b_h$. Suppose that the nodes of height h are MIN nodes and that A is one of them. Let $A_1, \ldots, A_b$ be the successors of A, and let $M(\ )$ mean "the minimax value of."

Then

$$p(h, l) = P(M(A) = 1 \mid L(A) = l)$$

$$= \frac{P(M(A) = 1, L(A) = l)}{P(L(A) = l)}$$

$$= \sum_{(l_i)} \frac{P(M(A_1) = 1, \ldots, M(A_b) = 1; L(A_1) = l_1, \ldots, L(A_b) = l_b)}{P(L(A) = l)}$$

$$= \sum_{(l_i)} \frac{P(M(A_1) = 1, L(A_1) = l_1)}{P(L(A_1) = l_1)} \cdots \frac{P(M(A_b) = 1, L(A_b) = l_b)}{P(L(A_b) = l_b)}$$

$$\times \frac{P(L(A_1) = l_1, \ldots, L(A_b) = l_b)}{P(L(A) = l)}$$

$$= \sum_{(l_i)} P(L(A_1) = l_1, \ldots, L(A_b) = l_b \mid L(A) = l)$$

$$\times p(h-1, l_1) \ldots p(h-1, l_b).$$

Therefore, from the conditional probability of $L(A_i)$ derived in the last section, we have

(B) $$p(h, l) = \sum_{(l_i)} \frac{\binom{b^{h-1}}{l_1} \cdots \binom{b^{h-1}}{l_b}}{\binom{b^h}{l}} \prod_{i=1}^{b} p((h-1), l_i).$$

The sum is over all b-tuples $(l_1, \ldots, l_b)$ such that $0 \leq l_i \leq b^{h-1}$ and $\sum_{i=1}^{b} l_i = l$.

For the other case, when the nodes of height h are MAX nodes, the following equation can be derived similarly:

(C) $$p(h, l) = 1 - \sum_{(l_i)} \frac{\binom{b^{h-1}}{l_1} \cdots \binom{b^{h-1}}{l_b}}{\binom{b^h}{l}} \prod_{i=1}^{b} (1 - p(h-1, l_i)),$$

where the sum is still over all b-tuples $(l_1, \ldots, l_b)$ such that $0 \leq l_i \leq b^{h-1}$ and $\sum_i l_i = l$. Therefore, any $p(h, l)$ can be calculated from the recurrence relations (B) and (C), with the initial conditions (A).

For an appropriate generating function of $p(h, l)$, equivalent but simpler recurrence relations can be derived as follows. For simplicity, MIN is assumed to be the last player. Then, the nodes of even height are MAX nodes and the nodes of odd height are MIN nodes. Let $G_h(h \geq 0)$ be the polynomial defined by

$$G_h(z) = \sum_l \binom{b^h}{l} p(h, l) z^l.$$

From the previous discussion of $p(h, l)$, it is easy to see that each coefficient of G_h is a non-negative integer. With this definition, the equation (B) becomes

$$G_{2h+1}(z)=(G_{2h}(z))^b,$$

and the equation (C) becomes

$$G_{2h}(z)=(1+z)^{b^{2h}}-((1+z)^{b^{2h-1}}-G_{2h-1}(z))^b.$$

Let $A(z)$ be defined by

$$A_h(z)=\frac{G_h(z)}{(1+z)^{b^h}};$$

then, A satisfies the following recurrence relation:

$$A_0(z)=\frac{z}{1+z},$$

$$A_{2h+1}=(A_{2h})^b,$$

$$A_{2h}=1-(1-A_{2h-1})^b.$$

Notice that these recurrence relations are exactly those for the probability that a node of height h is a forced win (Sect. 5.3). However, the equations above are independent of the initial probability, p, at the leaves. Furthermore, $p(h, l)$ is the ratio of the coefficient of z^l in the numerator to that in the denominator of A.

For P_2-games, the values $p(h, l)$ for $0 \le h \le 4$ are shown in Table 9.1, where MIN is still the last player.

Table 9.1. Some values of $p(h, l)$ for $b=2$

$16l/2^h$	$h=0$	$h=1$	$h=2$	$h=3$	$h=4$
0	0.0000	0.0000	0.0000	0.0000	0.0000
1					0.0000
2				0.0000	0.0000
3					0.0000
4				0.0000	0.0044
5					0.0220
6				0.0000	0.0649
7					0.1468
8		0.0000	0.3333	0.0571	0.2786
9					0.4602
10				0.2857	0.6683
11					0.8535
12			1.0000	0.7143	0.9648
13					1.0000
14				1.0000	1.0000
15					1.0000
16	1.0000	1.0000	1.0000	1.0000	1.0000

9.3   An Approximation of $p(h, l)$

When h is large, it is computationally expensive to compute $p(h, l)$ by repeatedly using the equations (B) and (C). Now we introduce a practical way to approximate $p(h, l)$. For simplicity, we still assume that MIN is the last player; that is, the nodes of even height are MAX nodes, and the nodes of odd height are MIN nodes.

For each integer $h \geq 0$, let the function e_h on the unit interval be recursively defined as follows. If $h = 0$, then

$$e_0(x) = x \quad (0 \leq x \leq 1).$$

If $n > 0$ is an integer, then

$$e_{2n}(x) = 1 - (1 - e_{2n-1}(x))^b, \text{ and}$$

$$e_{2n+1}(x) = (e_{2n}(x))^b \quad (0 \leq x \leq 1).$$

Note that the value $e_h(p)$ is exactly the value E_h (Sect. 5.3) and is the probability that a node of height h is a forced win (for MAX), where p is the probability that a leaf has the value 1. In approximating $p(h, l)$, notice that l/b^h, being an unbiased estimator of p, is first used as an estimate of p. Then, the value $e_h(l/b^h)$ is used to approximate $p(h, l)$. Although this approximated value is actually the probability that a node of height h is a forced win if each leaf has the value 1 with a probability of l/b^h, still $p(h, l)$ is the conditional probability that a node of height h is a forced win, given l ones under the node. Nau (1983a) used this approximation in his study of pathological phenomena.

There is as yet no theoretical study on the precision of this approximation. However, for the binary games ($b = 2$) and $1 \leq h \leq 8$, the maximum absolute difference of $p(h, l)$ and $e_h(l/2^h)$, shown in Table 9.2, decreases rapidly if the height, h, increases. That is, when the game tree is deep, this approximation seems to be practical.

Table 9.2. Maximum absolute error of the approximation of $p(h, l)$ for P_2-game models, where MIN is the last player

height	absolute error
$h = 1$	0.250
$h = 2$	0.191
$h = 3$	0.134
$h = 4$	0.085
$h = 5$	0.066
$h = 6$	0.045
$h = 7$	0.035
$h = 8$	0.025

10 Estimation of Minimax Values in G_d-Game Models

Since the structures of G_d-games are different from the structures of P_b-games, a G_d-game model has a different process for estimating minimax values based on the one-counter. In this chapter, we study the estimation of minimax values in G_d-game models.

Although the values 1 and 0 at leaves are randomly assigned in both P_b- and G_d-game models, the game graphs of the two models are very different. The game graph of a P_b-game model is a complete, uniform tree. All submodels at the same level of nodes are independent and occupy the same probabilistic space; therefore, the minimax values of nodes at the same level are independent and identically distributed (Sect. 5.3). However, in the game graph of a G_d-game model, any two consecutive nodes at the same level have exactly d common successors, and the submodels at the same level are highly interdependent, so that the minimax value of the root is equal to the minimax value of the middle node at each deeper level of the same type (MAX or MIN) as the root, excluding the leaves (Sect. 5.4). It is essential to take this property of G_d-games into account in developing a correct process for estimating minimax values.

Note that this same property also leads to an easy way of deriving actual minimax values from the values at the leaves; however, the main purpose of this chapter is to demonstrate that a correct estimating process may be very different from a conventional one. Specifically, it is shown that no back-up procedure is necessary in the estimating process introduced in this chapter; therefore, the research on heuristic game playing should not be focused solely on looking for new back-up processes.

For the one-counter, the corresponding heuristic searches in both a P_b-game model and a G_d-game model also have an essential difference. In a P_b-game model, the one-counter always returns more precise information at a deeper level in the game trees; therefore, we search for the heuristic information at only the search-frontier level. And in this case, a back-up process (using the product-propagation rules) is necessary in the estimation of minimax values.

However, as we have illustrated (Sect. 6.1.2), the one-counter does not necessarily return more precise information at a deeper level in a G_d-game model. Therefore, the estimation of minimax values based on the heuristic information at only search-frontier nodes is not necessarily as good as the estimation based on the heuristic information at all search nodes from the root down to the search frontier.

In order to improve the estimation at a deeper level, we need to search for more precise heuristic information at a deeper level; that is, we need to accumulate the heuristic information at all shallower levels.

Before we estimate minimax values in a G_d-game model, we first introduce a new process of accumulating the heuristic information; this process is equivalent to but simpler than the one-counter. To find the number of 1's in nodes, it is not necessary to count the number of 1's at each position because, in common successors, many squares in the board configuration would be uselessly searched many times. Given the total number of 1's in the original board configuration, only new squares need to be searched at each step. Through use of the special structure of the game graph, the number of 1's in certain positions can be easily derived.

Based on the cumulative heuristic information collected by this new process, the estimation of the minimax value of the root can be reduced to the estimation of the minimax value of only one node, given the number of 1's in this node. As mentioned above, no back-up procedure is necessary.

Both the process of accumulating heuristic information and the method of estimating minimax values are discussed in detail, first for G_1-game models. Then the results can be easily extended to general G_d-game models. We still assume that MIN is the last player. If MAX is the last player, we can easily modify the results by interchanging 1 and 0 at the leaves so that the roles of MAX and MIN can be reversed.

10.1   Estimation in G_1-Game Models

Consider the G_1-game tree in Fig. 10.1, where nodes are not labeled as belonging to MAX or MIN. Let the letters a, b, and so on, denote the search information, that is, the number of ones in each corresponding board configuration. The search starts from the root, where the total number of 1's on the playing board is found to be a; we assume that this total number of 1's in the original board configuration is given.

When the next level is searched, b and c are returned. From these three values, a, b, and c, the value x_1 at the left end and y_1 at the right end of the root can be derived as follows:

$$x_1 = a - c,$$
$$y_1 = a - b.$$

Thus, the search at the first level (the sons of the root) is reduced to the search for just the values x_1 and y_1 at both ends of the playing board, which has the form:

$$(x_1 \ldots y_1),$$

where the total number of ones is already given. Since the remaining part of the playing board (excluding x_1 and y_1) is the middle node at the second level, the value

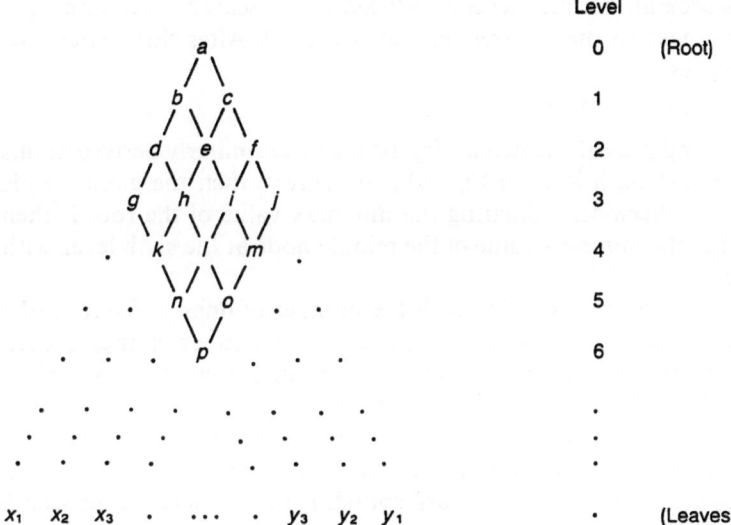

Fig. 10.1. A heuristic search in a G_1-game tree, where $a, b, \ldots,$ and p are the numbers of ones at the search nodes, and $x_1, x_2, x_3, y_1, y_2,$ and y_3 are values at the search squares

e of this node can also be derived:

$$e = a - x_1 - y_1$$
$$= b + c - a.$$

The search for values d and f at the second level is now reduced to the search for values x_2 and y_2, which are next to x_1 and y_1, respectively:

$$x_2 = c - f,$$
$$y_2 = b - d.$$

Similarly, the values h and i of the two middle nodes at the third level and the value l of the middle node at the fourth level can be derived from the previously found values:

$$h = d + e - b,$$
$$i = e + f - c,$$
$$l = h + i - e$$
$$= a - x_1 - y_1 - x_2 - y_2.$$

The minimax value of the root is equal to the minimax value of the middle node at the fourth level (Sect. 5.4); therefore, estimating the minimax value of the root, given the search information at the second level, is reduced to estimating the minimax value of this middle node at the fourth level, given l ones in the node, where l is derived as above.

Similarly, the search at the third level is reduced to the search from both sides for the two squares next to the squares already searched. After this search, the playing board becomes:

$$(x_1 \, x_2 \, x_3 \, \ldots \, y_3 \, y_2 \, y_1).$$

The values $k, m, n, o,$ and p, as illustrated in Fig. 10.1, can be similarly derived at this step. If the nodes at the sixth level in Fig. 10.1 are leaves, then the game tree is completely searched; otherwise, estimating the minimax value of the root is then reduced to estimating the minimax value of the middle node at the sixth level, with p ones in the node.

It has now been shown that, starting with the number of ones in the root (the whole playing board), the heuristic search that counts the number of ones in each node is simplified to a process searching only two new squares at each level. If we search $s \, (\geq 1)$ levels from the root, then we can calculate the number of ones in the middle node of level $2s$ (provided that the tree has more than $2s$ levels). The value of this node is the total number of ones on the remaining unsearched squares. We call this value the *search value*, and we call the corresponding node at level $2s$ the *search node* relative to level s.

Furthermore, estimating the minimax value of the root is reduced to estimating the minimax value of the search node, given the search value. Notice that there is no back-up process necessary in this estimation. If the height of the root is h, then the height of the search node relative to level s is $h - 2s$. This simplified search is summarized in Table 10.1.

Table 10.1. Summarized heuristic search in a G_1-game model with the initial value a, in which a square from each end is searched at each level

Search level	Board configuration	Search value	Height of search node
0	$(* * *)$	$v := a$	h
1	$(x_1 * * * y_1)$	$v := v - x_1 - y_1$	$h - 2$
2	$(x_1 \, x_2 * * * y_2 \, y_1)$	$v := v - x_2 - y_2$	$h - 4$
⋮	⋮	⋮	⋮
s	$(x_1 \ldots x_s *** y_s \ldots y_1)$	$v := v - x_s - y_s$	$h - 2s$

The problem of estimating the minimax value of the root in a G_1-game graph is now reduced to the problem of estimating the minimax value of a node N, given the number of 1's in the corresponding position. Let h be the height of N and l the number of 1's. Then the total number of squares in the board corresponding to N is $h + 1$. Consider the game graph rooted at N, and note that MIN is the last player.

Consider the first case when N is a MIN node; that is, h is odd. From Theorem 5.1, the minimax value of N is equal to the minimax value of the middle node N_1 of height 1. The minimax value of N_1 is 1 if and only if both squares of N_1 have the value 1. The other $l - 2$ ones may be arbitrarily distributed in other $h + 1 - 2 \, (= h - 1)$ remaining squares. Therefore, the number of all possible board configurations like this is the binomial coefficient:

$$\binom{h-1}{l-2}.$$

On the other hand, the number of all possible board configurations with l ones in the $h+1$ squares is

$$\binom{h+1}{l}.$$

Since all configurations are equally likely, the conditional probability that N is a forced win, given l ones is

(I′) $$p(h, l) = \frac{\binom{h-1}{l-2}}{\binom{h+1}{l}}, \quad h(>0) \text{ is odd}.$$

Second, let N be a MAX node; that is, h is even. In this case, the minimax value of N is equal to the minimax value of the middle node N_2 of height 2. Note that N_2 is a forced win if and only if the corresponding board configuration of N_2 is one of these two cases (1 1 *) or (0 1 1). In the first case, the third square of N_2 may have either 1 or 0; that is, the other $l-2$ ones may be arbitrarily distributed in the $h-1$ remaining squares. Therefore, the number of all such board configurations of N_2 is

$$\binom{h-1}{l-2}.$$

Similarly, the number of all possible board configurations of N_2 for the second case is

$$\binom{h-2}{l-2}.$$

Finally, we derive the conditional probability that N is a forced win, given l ones in the board:

(II′) $$p(h, l) = \frac{\binom{h-1}{l-2} + \binom{h-2}{l-2}}{\binom{h+1}{l}}, \quad h(>0) \text{ is even}.$$

Since $p(h, l)$ has a simple closed form (I′ or II′), its value can be easily calculated for any case. Some $p(h, l)$ are displayed in Table 10.2.

10.2   Estimation in G_d-Game Models

Now we extend the results of the last section to general G_d-game models. The minimax value of the root of a G_d-game graph is, like that in a G_1-game graph,

Table 10.2. The values of $p(h, l)$ for a G_1-game model and for $h = 2, \ldots, 13$

h	2	3	4	5	6	7	8	9	10	11	12	13
l												
1	0.000	0.000	0.000	0.000	0.000	0.000	0.000	0.000	0.000	0.000	0.000	0.000
2	0.667	0.167	0.200	0.067	0.095	0.036	0.056	0.022	0.036	0.015	0.026	0.011
3	1.000	0.500	0.500	0.200	0.257	0.107	0.155	0.067	0.103	0.045	0.073	0.033
4		1.000	0.800	0.400	0.457	0.214	0.286	0.133	0.194	0.091	0.140	0.066
5			1.000	0.667	0.667	0.357	0.437	0.222	0.303	0.152	0.221	0.110
6				1.000	0.857	0.536	0.595	0.333	0.424	0.227	0.315	0.165
7					1.000	0.750	0.750	0.467	0.552	0.318	0.416	0.231
8						1.000	0.889	0.622	0.679	0.424	0.522	0.308
9							1.000	0.800	0.800	0.545	0.629	0.396
10								1.000	0.909	0.682	0.734	0.495
11									1.000	0.833	0.833	0.604
12										1.000	0.923	0.725
13											1.000	0.857
14												1.000

equal to the minimax value of the middle node at each level of the same type as the root. Therefore, the heuristic information at these middle nodes also determines the estimate of the minimax value of the root. In the following discussion, we first consider the derivation of relevant heuristic information at the middle nodes.

Consider a G_d-game model. Let the height of the root of the corresponding G_d-game graph be h; then, there are a total of $hd + 1$ squares in the original playing board. Since the one-counter returns the number of l's in this initial position by searching the root, we also assume that for each game in the model the number of l's is previously given, and is denoted by a. In other words, the heuristic information at the root, denoting the number of l's in the initial playing board, is given for each chosen game.

Consider the heuristic search at the successors of the root, illustrated in Fig. 10.2. The one-counter returns at these successors the values $b_1, \ldots, b_{d+1}$, each of which represents the number of l's in each corresponding position. Let C be the only common son of the successors, that is, the middle node at the next level. Let $x_1, \ldots, x_d$ be the values of the first d squares from the left end, and let $y_1, \ldots, y_d$ be the values of the first d squares from the right end of the initial position. Then in the node C, the number of l's, denoted by c, can be derived as follows:

$$c = b_1 + b_{d+1} - a$$

$$= a - x_1 - x_2 - \ldots - x_d - y_1 - y_2 - \ldots - y_d.$$

The number of l's in any other deeper node, however, cannot be derived from only these b_i's.

Since the minimax value of the root is equal to the minimax value of the middle node C at level 2, estimating the minimax value of the root, based on a and the b_i's, is then equivalent to estimating the minimax value of the node C based on the

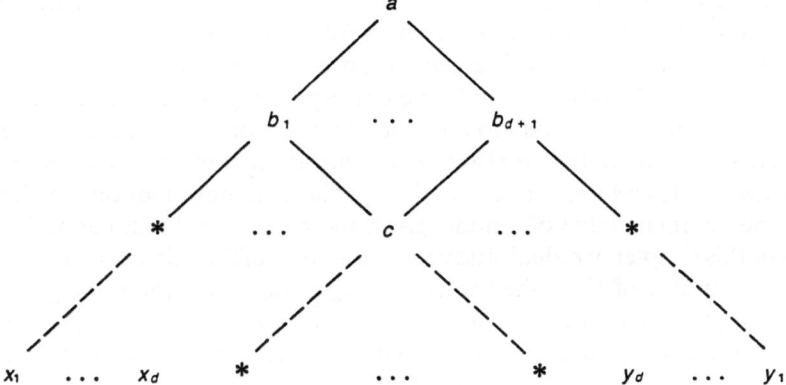

Fig. 10.2. A heuristic search in a G_d-game model at level 1 with the initial value a at the root. The first d squares from both ends are searched, and the value of the middle node at the next level is $c = a - x_1 - \ldots - x_d - y_1 - \ldots - y_d$

derived value $c (= b_1 + b_{d+1} - a)$. Therefore, the heuristic search of level 1 is reduced to the search for c, the number of 1's in the middle node of level 2. The value c is found by subtracting from a the number of 1's in the first d squares from each end of the board; that is,

$$c = a - x_1 - x_2 - \ldots - x_d - y_1 - y_2 - \ldots - y_d.$$

The middle node C at level 2 and the returned value c are, again as in G_1-game models, called the *search node* and the *search value* of level 1, respectively. To search deeper, we simply repeat the process above; that is, at each deeper level, we search another d new squares from each end and subtract from the old search value the number of 1's in these $2d$ newly searched squares. The new search value is the number of 1's in the remaining unsearched squares, which form the new search node, that is, the middle node at the next second deeper level than the old search node. This heuristic search is summarized in Table 10.3.

Table 10.3. Summarized heuristic search in a G_d-game model with the initial value a, in which d squares from each end are searched at each level

Search level	Board configuration	Search value	Height of search node
0	$(***)$	$v := a$	h
1	$(x_1 \ldots x_d *** y_d \ldots y_1)$	$v := v - x_1 \ - \ldots - x_d$ $- y_1 \ - \ldots - y_d$	$h - 2$
2	$(x_1 \ldots x_{2d} *** y_{2d} \ldots y_1)$	$v := v - x_{d+1} - \ldots - x_{2d}$ $- y_{d+1} - \ldots - y_{2d}$	$h - 4$
$\vdots$	$\vdots$	$\vdots$	$\vdots$
s	$(x_1 \ldots x_{sd} *** y_{sd} \ldots y_1)$	$v := v - x_{(s-1)d} - \ldots - x_{sd}$ $- y_{(s-1)d} \ - \ldots - y_{sd}$	$h - 2s$

The estimation of the minimax value of the root, based on the heuristic information returned by the one-counter searching from the root down to a level of the game graph, is now reduced to the estimation of the minimax value of the corresponding search node, based on only the corresponding search value. Since the minimax value of the root is equal to the minimax value of the search node, the estimate derived at the search node is also the estimate at the root, and no back-up process is necessary. Therefore, our estimation problem is now reduced to the estimation of the minimax value of a node, given the number of 1's in the node.

In the rest of this chapter, we shall study the estimation of the minimax value of a node, given the number of 1's in the corresponding position. Consider the game with the game graph rooted at this node. Suppose that the height of this node is h and that the number of 1's in the playing board is l; then, the whole board has $hd + 1$ squares. The estimate is also denoted by

$$p(h, l),$$

which is the conditional probability that the node is a forced win, given l ones in the corresponding game position of the node. It is still assumed that MIN is the last player; that is, the nodes of odd height are MIN nodes, and the nodes of even height are MAX nodes. Each is considered separately.

Case 1. A MIN Node

If the root of the game graph is a MIN node, then its minimax value is equal to the minimax value of the middle node of height 1 (Theorem 5.2). Therefore, the estimate $p(h, l)$ of the root is also the conditional probability that the middle node of height 1 is a forced win, given l ones in the initial playing board.

Being a MIN node, a node of height 1 is a forced win (for MAX) if and only if all $d + 1$ squares in the corresponding position have the value 1; these $d + 1$ squares are the middle $d + 1$ squares in the initial playing board. Therefore, the root is a forced win if and only if the middle $d + 1$ squares have $d + 1$ ones. Since the other $l - d - 1$ ones can be arbitrarily distributed in the remaining $hd + 1 - d - 1 (= (h-1)d)$ squares, the number of all possible such board configurations is

$$\binom{(h-1)d}{l-d-1}.$$

Since all possible board configurations are equally likely and since the number of all such board configurations is

$$\binom{hd+1}{l},$$

we derive the estimate as follows:

(I) $$p(h, l) = \frac{\binom{(h-1)d}{l-d-1}}{\binom{hd+1}{l}}, \quad h(>0) \text{ is odd.}$$

Case 2. A MAX Node

In this case, the minimax value of the root is equal to the minimax value of the middle node of height 2, the middle $2d+1$ squares. And the estimate, $p(h, l)$, is then the conditional probability that this middle node of height 2 is a forced win (for MAX), given l ones in the initial playing board.

Since it is a MAX node, the middle node of height 2 is a forced win if and only if at least one of its successors, being of height 1, is a forced win, i.e., if in its board configuration with $2d+1$ squares, there exist $d+1$ consecutive ones. Now assume that the middle node of height 2, denoted by N, is a forced win. We want to find the number of possible board configurations.

Since N has at least one successor that is a forced win, let the ith ($1 \leq i \leq d+1$) successor be the first forced win from the left. If $i=1$, then the first $d+1$ squares of N have the value 1, as illustrated in Fig. 10.3, and the other $l-d-1$ ones can be arbitrarily arranged in the remaining $(h-1)d$ $(=hd+1-d-1)$ squares. Therefore, the number of all possible board configurations for $i=1$ is

$$\binom{(h-1)d}{l-d-1}.$$

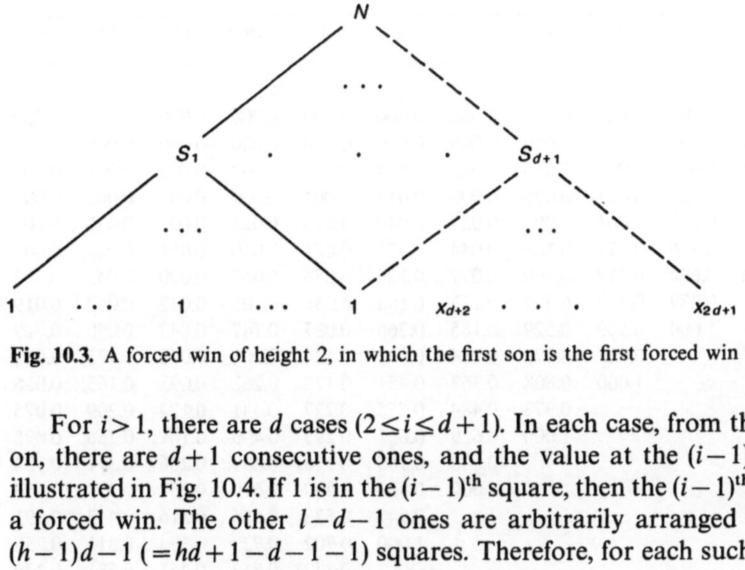

Fig. 10.3. A forced win of height 2, in which the first son is the first forced win from the left

For $i>1$, there are d cases ($2 \leq i \leq d+1$). In each case, from the ith square of N on, there are $d+1$ consecutive ones, and the value at the $(i-1)^{th}$ square is 0, as illustrated in Fig. 10.4. If 1 is in the $(i-1)^{th}$ square, then the $(i-1)^{th}$ successor of N is a forced win. The other $l-d-1$ ones are arbitrarily arranged in the remaining $(h-1)d-1$ $(=hd+1-d-1-1)$ squares. Therefore, for each such i, the number of all possible board configurations is

$$\binom{(h-1)d-1}{l-d-1}.$$

For the above analysis, the number of all possible board configurations in which N is a forced win is

$$\binom{(h-1)d}{l-d-1} + d\binom{(h-1)d-1}{l-d-1};$$

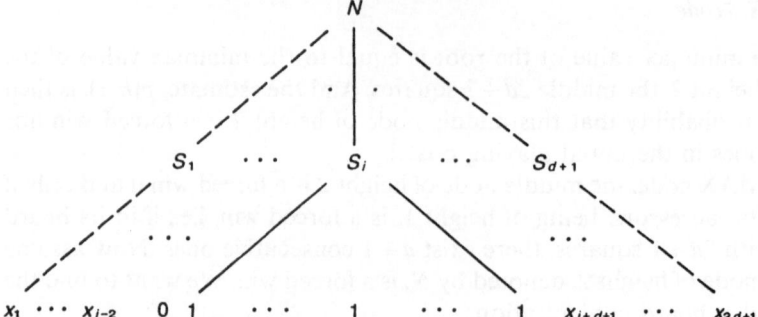

Fig. 10.4. A forced win of height 2, in which the ith son is the first forced win from the left $(1 < i)$

Table 10.4. The values of $p(h, l)$ for a G_2-game model with $h = 2, \ldots, 13$

h	2	3	4	5	6	7	8	9	10	11	12	13
l												
1	0.000	0.000	0.000	0.000	0.000	0.000	0.000	0.000	0.000	0.000	0.000	0.000
2	0.000	0.000	0.000	0.000	0.000	0.000	0.000	0.000	0.000	0.000	0.000	0.000
3	0.300	0.029	0.036	0.006	0.010	0.002	0.004	0.001	0.002	0.001	0.001	0.000
4	0.800	0.114	0.127	0.024	0.039	0.009	0.017	0.004	0.009	0.002	0.005	0.001
5	1.000	0.286	0.278	0.061	0.091	0.022	0.040	0.010	0.021	0.006	0.012	0.003
6		0.571	0.476	0.121	0.168	0.044	0.076	0.021	0.040	0.011	0.024	0.007
7		1.000	0.694	0.212	0.269	0.077	0.125	0.036	0.067	0.020	0.040	0.012
8			0.889	0.339	0.392	0.123	0.188	0.058	0.103	0.032	0.062	0.019
9			1.000	0.509	0.529	0.185	0.265	0.087	0.147	0.047	0.090	0.029
10				0.727	0.671	0.264	0.353	0.124	0.201	0.068	0.123	0.041
11				1.000	0.808	0.363	0.451	0.170	0.262	0.093	0.163	0.056
12					0.923	0.484	0.555	0.227	0.331	0.124	0.209	0.075
13					1.000	0.629	0.661	0.295	0.406	0.161	0.260	0.098
14						0.800	0.765	0.376	0.487	0.206	0.317	0.124
15						1.000	0.860	0.470	0.570	0.257	0.378	0.156
16							0.941	0.578	0.655	0.316	0.443	0.191
17							1.000	0.702	0.739	0.384	0.511	0.232
18								0.842	0.818	0.461	0.581	0.279
19								1.000	0.890	0.547	0.651	0.331
20									0.952	0.644	0.721	0.390
21									1.000	0.751	0.789	0.455
22										0.870	0.852	0.526
23										1.000	0.910	0.605
24											0.960	0.692
25											1.000	0.786
26												0.889
27												1.000

therefore, the estimate at the root is

(II) $$p(h, l) = \frac{\binom{(h-1)d}{l-d-1} + d\binom{(h-1)d-1}{l-d-1}}{\binom{hd+1}{l}}, \quad h(>0) \text{ is even.}$$

If we set $d=1$ in the formulas (I) and (II), we obtain, as special cases, the formulas (I') and (II') previously derived in Sect. 10.1. Since both formulas are in closed form, the estimate $p(h, l)$ can be easily calculated for any h and l in general G_d-game models. Some estimates for $d=2$ are displayed in Table 10.4.

The searching process and the method of estimating minimax values derived in this chapter for G_d-game models are quite different from conventional approaches. Based on the structure of G_d-game models, a special process can be designed to search for the heuristic information at only those nodes relevant to the minimax value of the root. In estimating the minimax value of the root, it is sufficient to estimate the minimax value of the search node, and no back-up process is involved at all.

The aim of this chapter has been to demonstrate, by using simple G_d-game models, that the structure of an actual search process may be very different for different games. The conventional approach, using a static evaluation function and a back-up process, is not always applicable. In developing a heuristic search process for a problem, it is important to take into account the structures of the problem and the properties of relevant heuristic information.

11 Conclusions

The mathematical theory of heuristic information introduced in this book demonstrates the fact that concepts in heuristic game-tree searches, such as heuristic information, node strength, node estimation, and decision making, can be formulated in mathematical terms. In a heuristic game-tree search, if the information is accumulated at all of the nodes in the search tree and if the node strength of each possible next move is precisely estimated, then the decision quality is improved with a larger search tree. Accumulating information at all nodes in the search tree guarantees the increase of heuristic information in a deeper search. The increased information then upgrades node-strength estimates and hence improves decision quality.

Two arguments can explain pathology found in recent experiments on conventional game-tree searches. The pathology is the phenomenon that in some game models reaching deeper consistently degrades the quality of a decision. First, the conventional minimaxing process backs up values at search-tip nodes, and only the information at search-tip nodes is used. The back-up process fails to save information from further up in the tree, which could have been used to break ties or possibly to prevent pathology in some cases. Second, minimaxing in general is not theoretically correct in estimating nodes. Since estimating a node strength is a mathematical problem in our abstract model, the method of doing this should depend on each individual case. For example, the product-propagation rules are proved to be suitable for independent game models (Chap. 8). This theoretical conclusion concurs with an experimental result indicating that the product-propagation rules prevent pathology in P_b-game models (Nau 1983a).

The pathological phenomenon is not observed in common game playing. Computer programs playing common games such as chess and checkers improve their performances with increasing search depth. There are several arguments about the difference between common games and the game models in pathology experiments (Pearl 1984, Beal 1982, Bratko and Gams 1982).

Since P_b-game models are pathological (Nau 1983a, Pearl 1983), it is argued that the assumption of independence may not fit common games such as chess and checkers. Independent game models such as P_b-game models have two special characteristics. First, each game graph is a uniform game tree which grows exponentially. No two game positions in the game tree are identical. By contrast, different paths in the game graph of a common game often result in a same game position. Second, minimax values of nodes at the same level are completely independent. Even neighboring nodes have independent WIN-LOSS status. However, in common games, the strength of a board position changes in an incremental manner so that closely related nodes such as sibling nodes have closely related minimax values.

Consider the other extreme cases, G_d-game models. The minimaxing process is not pathological in G_d-game models (Nau 1983b). The unique characteristic of G_d-game models is the extreme dependence of nodes in the game graphs. Two features stem from this characteristic. First, there are many identical positions in a game tree and the merged game graph does not grow exponentially in size (Chap. 5). Second, the nodes at the same level are highly correlated so that minimax values of nodes have simple relationships with each other (Theorem 5.2). Based on those relationships, simple algorithms can be designed to find the minimax value of the root with almost constant effort, independent of the size of the graph. Therefore, an idealized game playing with actual minimax values is possible. Even in a heuristic search, although minimaxing is not correct in estimating nodes (Chap. 10), it is still good enough to avoid pathology.

P_b-game models and G_d-game models are two extreme cases: completely independent and totally dependent. Common games are somewhat between these two extreme cases; some dependence exists between neighboring nodes but the game graphs still grow exponentially in size. In a simulation study, Nau (1981) used the board-splitting game (as a P_2-game); however, the initial assignment of WIN-LOSS to the individual cells was governed by a process that maintained a strong correlation between neighboring cells. Such dependence causes the pathology to disappear. This suggests that the incremental change in node strength in games such as chess and checkers is one of the reasons why such games are not pathological.

The notion of node strength is only heuristic and cannot be precisely formulated in the study of game-tree searches. Although minimax values represent the strength of nodes in the idealized game theory, they are no longer useful in most actual games when both players are using heuristic methods. In the abstract model presented in this book, it is assumed that a node strength is always a previously given random variable. If both players' strategies are given from a node, then the "strength" of this node is the "game value" relative to these given strategies (Sect. 5.2). The minimax value of a node is a special game value relative to "theoretically optimal" strategies.

In a case where the opponent's strategy is not known, the problem of choosing a random variable to represent the "strength" of a node is still open. Based upon previous moves, the opponent's behavior can sometimes be predicted to a certain degree. How to incorporate such a predicted strategy into a decision process is an interesting problem for further study.

Appendix

Mathematical Theorems

This appendix introduces some theorems required in the discussion in Chap. 5 of the asymptotic behavior of minimax values in P_b-game models. Throughout this appendix, b is an integer greater than 1.

Theorem 1. In the interval $(0, 1)$ there is exactly one root of the equation

(A) $$x^b + x - 1 = 0.$$

This root is denoted by W_b.

Proof. Consider the function $f(x) = x^b + x - 1$. Since f is strictly monotonically increasing in the interval $[0, 1]$, and since $f(0) = -1$ and $f(1) = 1$, there is exactly one zero in $(0, 1)$. □

Theorem 2. The sequence $\{W_b\}$ with all possible b's is a strictly increasing sequence and
$$\lim_{b \to \infty} W_b = 1.$$

Proof. Note that $0 < W_b < 1$ and that $W_b = 1 - W_b^b$ for each $b > 1$. Let $b < c$. The assertion $W_b \geq W_c$ would lead to the following contradiction:
$$W_b = 1 - (W_b)^b \leq 1 - (W_c)^b < 1 - (W_c)^c = W_c.$$

Therefore, $\{W_b\}$ is strictly increasing. Let
$$\lim_{b \to \infty} W_b = W.$$

Then $W_b < W \leq 1$ for all b. If $W < 1$, then $W_b^b < W^b$ for all b and
$$\lim_{b \to \infty} W_b^b = 0.$$

This leads to another contradiction:
$$\lim_{b \to \infty} W_b = \lim_{b \to \infty} (1 - W_b^b) = 1 - 0 = 1. \qquad \square$$

Theorem 3. W_b is also a root of the equation:

(B) $$1 - (1 - x^b)^b = x.$$

Proof. From $W_b^b + W_b - 1 = 0$, we have $1 - W_b^b = W_b$ and
$$1 - (1 - W_b^b)^b = 1 - W_b^b = W_b. \qquad \square$$

Theorem 4. W_b is the only root of the equation (B) in the interval (0, 1).

Proof. Consider the function $y = 1 - (1 - x^b)^b - x$. Then y has 3 different zeros, 0, W_b, and 1. If there were another zero in (0, 1) then the derivative of y' would have at least three zeros, and y'' would have at least two zeros in (0, 1). However, since

$$y'' = b^2(b-1)x^{b-2}(1-x^b)^{b-2}(1-(b+1)x^b),$$

y'' has only one zero in (0, 1). Therefore, y has only one zero in (0, 1). $\square$

Theorem 5. The following two inequalities hold:

$$1 - (1 - x^b)^b < x \quad \text{for} \quad 0 < x < W_b$$

and

$$1 - (1 - x^b)^b > x \quad \text{for} \quad W_b < x < 1.$$

Proof. Let

$$y = 1 - (1 - x^b)^b - x.$$

Since

$$y' = b^2 x^{b-1}(1 - x^b)^{b-1} - 1,$$

$y'(0)$ and $y'(1)$ are negative, and thus y is decreasing at 0 and 1. The asserted inequalities now come from the fact that y is zero only at 0, W_b and 1. $\square$

Let p be any value in the interval [0, 1]. For any integer $b > 1$, define P_n recursively as follows:

$$P_0 = p,$$

$$P_n = 1 - (1 - P_{n-1}{}^b)^b \quad \text{for} \quad n > 0.$$

Theorem 6. If $P = W_b$, then $P_n = W_b$ for all n; otherwise,

$$\lim_{n \to \infty} P_n = 0 \quad \text{for} \quad 0 \le p < W_b,$$

and

$$\lim_{n \to \infty} P_n = 1 \quad \text{for} \quad W_b < p \le 1.$$

Proof. The first assertion in this theorem comes directly from Theorem 3. Consider next the second assertion. For the cases when $p = 0$ or 1, the conclusions come immediately from the definition. Now suppose $0 < p < W_b$. Consider the function

$$f(x) = 1 - (1 - x^b)^b.$$

Then

$$P_n = f(P_{n-1}),$$

and each P_n is positive. From Theorem 5, P_n is strictly decreasing. Let

$$W = \lim_{n \to \infty} P_n.$$

Then $0 \le W < W_b$. W is a fixed point of f, and from Theorem 4 we have $W = 0$. The assertion for the other case (i.e., $W_b < p < 1$) can be proved similarly. $\square$

References

Barr A, Feigenbaum EA (eds) (1981) The handbook of artificial intelligence, vol 1. William Kauffmann, Los Altos, CA

Baudet GM (1978) On the branching factor of the alpha-beta pruning algorithm. Artificial Intelligence 10(2): 173–199

Beal D (1980) An analysis of minimax. In: Clarke MRB (ed) Advances in computer chess 2. Edinburgh University Press, Edinburgh, pp 103–109

Beal D (1982) Benefits of minimax search. In: Clarke MRB (ed) Advances in computer chess 3. Pergamon, Oxford, pp 17–24

Bratko I, Gams M (1982) Error analysis of the minimax principle. In: Clarke MRB (ed) Advances in computer chess 3. Pergamon, Oxford, pp 1–15

Chung KL (1974) A course in probability theory, 2nd ed Academic Press, New York

Dresher M (1981) The mathematics of games of strategy: theory and application. Dover Publications, New York

Knuth DE, Moore RW (1975) An analysis of alpha-beta pruning. Artificial Intelligence 6(4): 293–326

Nau DS (1980) Decision quality as a function of search depth of game trees. TR-866, Computer Science Department, University of Maryland

Nau DS (1981) An investigation of the causes of pathology in games. TR-999, Computer Science Department, University of Maryland

Nau DS (1983a) Pathology on game trees revisited, and an alternative to minimaxing. Artificial Intelligence 21: 221–244

Nau DS (1983b) Game graph structure and its influence on pathology. TR-1246, Computer Science Department, University of Maryland

Nau DS, Purdom P, Tzeng C-H (1986) An evaluation of two alternatives to minimax. In: Kanaland LN, Lemmer JE (eds) Uncertainty in artificial intelligence. North-Holland, Amsterdam, pp 505–509

Nilsson N (1980) Principles of artificial intelligence. Tioga, Palo Alto, CA

Pearl J (1980) Asymptotic properties of minimal trees and game-searching procedures. Artificial Intelligence 14: 113–138

Pearl J (1981) Heuristic search theory: a survey of recent results. In: Proc IJCAI 7, Vancouver, British Columbia, Canada, pp 24–28

Pearl J (1983) On the nature of pathology in game searching. Artificial Intelligence 20: 427–453

Pearl J (1984) Heuristics: intelligent search strategies for computer problem solving. Addison-Wesley, Reading, MA

Pohl I (1970) First results on the effect of error in heuristic search. In: Meltzer B, Michie D (eds) Machine intelligence 5. American Elsevier, New York, pp 219–236

Reibman AL, Ballard BW (1983a) Non-minimax search strategies for use against fallible opponents. In: Proceedings of the national conference on artificial intelligence AAAI-83. William Kaufmann, Los Altos, CA, pp 338–34

Reibman AL, Ballard BW (1983b) The performance of a non-minimax search strategy in games with imperfect players. CS-1983-17, Duck University, Durham, NC

Samuel AL (1959) Some studies in machine learning using the game of checkers. IBM Journal R & D 3: 211–229

Slagle R, Dixon J (1970) Experiments with M & N tree-searching programs. Communication of ACM 13: 147–153

Tzeng C-H (1984) A mathematical model of heuristic game playing. In: Laubsch J (ed) GWAI-84. Springer-Verlag, Berlin, Heidelberg pp 209–218

Tzeng C-H, Purdom P (1983) A theory of game trees. In: Proceedings of the national conferences on artificial intelligence AAAI-83. William Kaufmann, Los Altos, CA, pp 416–419

Tzeng C-H, Purdom P (1986) Estimation of minimax values. In: Ras ZW, Zemankova M (eds) Proceedings SIGART international symposium on methodologies for intelligent systems. ACM, New York, pp 174–182

von Neumann J, Morgenstern M (1946) Theory of games and economic behavior. Princeton University Press, Princeton, NJ

Subject Index

Additivity 32
  countable 32
  finite 32
Adversary 2
Almost everywhere 38
Alpha-beta procedure 13, 14, 16, 20, 59
AND 48, 49, 50
Asymptotic behavior 46
Average 34
Average propagation 26

Back-up process 5, 18, 20
Backgammon 7
Ballard 25, 26
Barr 21
Baudet 17
Baye's rule 37
Bayesian statistics 37
Beal 4, 22, 98
Binomial distribution 82
BLACK 10
Bonus function 25
Boolean operators 48
Borel field 29, 30, 65
  product 35
  total 29
  trivial 29, 38
Borel set
  linear 33
Branching factor 45, 81
Bratko 98
Bridge 7

Checker 98
Chess 7, 98
Chung 28, 34
City-block distance 3, 4
Combinatorial explosion 3
Conditional expectation 37, 38
Conditional probability 36, 82, 91, 93
Control strategy 3
Control system 1
Countable additivity 32

Cutoff
  alpha 15
  beta 15
  lower 15
  upper 15

Decision behavior 40
Decision making 4, 64, 68
Decision model 68, 69
Decision problem 68
Decision quality 68, 70, 71, 72
Decision strategy 68
De Morgan's law 29
Discrete 32
Distance
  city-block 3
  Manhattan 3
Dixon 24, 25
Dresher 13, 41

8-Puzzle 1, 2, 4
Equation 54
  system 54
Estimation 5
Estimator 65
  A- 65
  B- 67
  more precise 67
Evaluation function
  static 3, 4, 18, 19, 41, 52, 58
Event 29, 58
Everywhere 38
Expectation 33
  conditional 37, 38, 65

Face-value principle 20, 22, 69
Fallible 25
Feigenbaum 21
Finite intersection 29
Forced loss 78, 79
Forced win 22, 66, 78, 79, 82, 91, 94, 95
Fubini's Theorem 77
Function
  measurable 33

Gambling 39
Game
    finite 5
    G_1- 10, 20, 47, 56
    G_2- 47
    G_d- 5, 10, 20, 47
    P_2- 11, 20, 52
    P_b- 5, 20, 22
    payoff 13
    perfect information 5, 7, 18
    player 18
    T- 41
    two-person 5, 18
    WIN-LOSS 5, 66, 73, 78
    zero-sum 5, 18
Game graph 9, 28, 40, 67
    G_1- 11, 48
    G_2- 47
    G_d- 47
    height 9, 40, 48, 49
Game model
    product 73, 74
    G_1- 52, 56, 88
    G_d- 46, 47, 87, 91, 99
    P_2- 52
    P_b- 45, 81, 98
Game set 54, 56
Game space
    P_2- 28
Game theory 45
Game tree 8, 28, 40
    level 10, 41
    link 8
    nodes 9
    non-terminal node 8, 9
    predecessor 9
    root 8, 9, 40
    sons 8
    subtree 9
    successor 8, 9
    terminal node 9, 40
Game value 13, 41, 42, 44, 64
Game-tree search 4, 15, 52, 59
    heuristic 52, 59
    pathology 4, 18, 21, 22, 52, 69, 98
Gams 98
Global database 1
Goal 2
Graph
    product 73, 74
    search 2, 4

Height 9, 40, 45, 48, 49, 81
Heuristic information 3, 4, 5, 28, 51, 57, 58, 94, 98

cumulative 52, 53, 57, 60, 62
    local 75, 78
    product 75, 76, 77, 80
Heuristic search 3, 4, 5, 55, 59, 61, 65, 68
    information 60
    local 76
    method 3
    product 75, 76
HORIZONTAL 11

I.i.d. 45, 81
Inclusion 61
Independence 34, 73, 75
Indicator 66
Information
    complete 53, 59
    cumulative 53, 62
    partial 71
    trivial 59
Integrable 34
Integration 33

Kalah 25
Knuth 17

Last player 48, 88
Leaves 40
LEFT 8
Localization 75
LOSS 23

Manhattan distance 3
Martingale 5, 28, 36, 39, 67
MAX 9, 10, 40, 43, 44, 45, 46, 47, 64
Mean 33, 34, 45
Measurable set 31, 58
Measurable function 33
Measure 31
    product 35
MIN 9, 10, 40, 43, 44, 45, 46, 47, 64
Minimal cost 3
Minimax procedure 13, 14, 20, 59
Minimax value 5, 13, 18, 40, 44, 81, 87, 89, 94
Misplaced tile 3
Moore 17
Morgenstern 13
Move
    chance 7
    minimax optimal 14
    optimal 13
    personal 7

Nau   4, 11, 20, 22, 26, 27, 86, 98, 99
N-decision random variable   70, 71
N-decision-making   70, 71
Negmax   17
Nilsson   3, 19
Node
   LOSS   23
   MAX   9, 40, 44, 49, 50, 80, 95
   MIN   9, 40, 44, 49, 50, 80, 94
   search-tip   3, 19
   terminal   40
   WIN   23
Node strength   28, 64
Non-adversary   2
Non-comparable   61
NP-complete   5

One-counter   5, 52, 83, 88, 92
Optimal solution   3
OR   48, 49, 50

Partition   29
   finer   30
   proper refinement   30
   properly finer   30
   refinement   30
Pathological phenomenon   4, 18, 21, 22, 52, 69, 98
Payoff   9
   final   9
Pearl   3, 4, 17, 22, 24, 45, 73, 98
Performance quality   22
Playing path   42
Pohl   4
Poker   7
PP rules   80
PP-1   79
PP-2   79
Precise
   equally   55, 61
   more   55, 61
   non-comparable   61
   properly more   55, 61
Predicted strength   25
Probabilistic game model   5, 40, 41
Probabilistic game space   32
Probability measure   30
Probability space   30
   product   35
Problem
   graph-search   2
   NP-complete   3
   optimal   3

Problem state   2
Procedure
   *-MIN   25, 26
   alpha-beta   13, 14, 16, 20, 59
   M & N   24
   minimax   13, 14, 20, 59
   product-propagation   5, 23, 24, 73, 77, 80, 81, 83, 98
Product
   subgraph   73, 74
   game model   73, 74
   component   74, 75
Product measure   35
Product-propagation   5, 23, 24, 73, 77, 80, 81, 83, 98
Product set   35
Product space   35
Production rules   1, 2
Production system   1, 2
Purdom   26, 27, 66

Random function   58
Random variable   33, 65
   estimation   65
   i.i.d.,   45, 81
   independent   34, 73, 75
   mean   34, 45
Random vector   41
Recurrence relation   46, 83, 84, 85
Reibman   25, 26
RIGHT   8
Root   40
Round   42
   complete   42

Samuel   18
Search depth   21
Search event   60, 61
Search information   60, 61
Search node   61, 90, 93
Search value   90, 93
Singleton   59
Slagle   24, 25
Static value   19
Strategy   41
   heuristic   18
   minimax optimal   44, 45, 64, 68
   non-randomized   42
   randomized   43
Strength   64, 69
Subfield   29
Subgame   9
Successor   9

T-game 41
Tic-tac-toe 1, 2, 19
Tzeng 26, 27, 41, 66

VERTICAL 11
Visibility 22, 30, 61, 63
  improved 63

von Neumann 13

Weighted average 26
WHITE 10
WIN 23

Zero-counter 61